Island Voices

John C. Kennedy

Island Voices:

Fisheries and Community Survival in Northern Norway

Eburon Delft
2006

Printed on acid-free paper.

ISBN-10: 90-5972-103-9
ISBN-13: 978-90-5972-103-6

Eburon Academic Publishers
P.O. Box 2867
NL-2601 CW Delft
The Netherlands
Phone: +31-(0)15-2131484 / Fax: +31-(0)15-2146888
info@eburon.nl / www.eburon.nl

Cover design: Hermkens Design, Amsterdam
Cover photograph: Annie Giæver (1882-1956): *Women carrying fish from the Årviksand beach* (1920s). Courtesy of the Tromsø University Museum Photo Archive.

Contents

FIGURES AND TABLES

ACKNOWLEDGEMENTS

A number of individuals and institutions provided financial support necessary for the completion of this book. I travelled to Norway in 1995 under Memorial University's Bergen Exchange program, and thank those responsible at Memorial and the University of Bergen. Memorial University's Institute of Social and Economic Research (ISER), long supportive of comparative research relevant to Newfoundland and Labrador, provided a research grant that helped fund the 1995 fieldwork; I thank both Memorial and ISER. Dr. Svein Jentoft's request to the University of Tromsø's Norges Fiskerihøgskole (NFH – the Norwegian Fishery High School), secured support to assist with publication costs, and I thank Dr. Jentoft and NFH Director, Professor Petter Holm. Moreover, Professors Holm and Jentoft, as well as Pernille Jørgensen, of the Årviksand Bygdelag (council), and Rolf Egil Larsen, of Lauksletta, all applied to the Chairman (formannskapet) of Skjervøy Kommune for additional funding to help defray publication costs, and I thank the Kommune leaders for voting to approve such support. I very much appreciate the help of these people and the institutions.

A number of friends and colleagues in several locations helped the project in diverse ways. At Memorial, I acknowledge the inspiration and support of students and fellow faculty, including Jill Allison, Louis Chairamonte, Dianne Grant, Barbara Neis, Steve Noel, Monica Playfoot, Peter Sinclair, Michael Tilley, Jim Rice, and Gerry White. At Bergen, I thank our gracious hosts, Georg and Berit Henriksen. At Tromsø, I thank Bjørn Bjerkli, Ivar Bjørklund, Håvord Dahl Bratrein, Ottar Brox, Steinar Eliassen, Geir Finne, Siri Gerrard, Petter Holm, Ernst Isaksen, Svein and Greta Jentoft, Per Mathiesen, Anita Maurstad, Ragnar Nilsen, Gry Paulgaard, Sidsel Saugestad, Trond Truen, and Nils Aarsæther; at Skjervøy and on Arnøya, I thank Nils Alm, Tor Arnesen, Carl Johan and Margrete Bentsen, Halvor Eilertsen, Ottar and Unni Friis, Helge Guttormsen, Øyvind Isaksen, Marianne Jakobsen, Pernille and Jan Erik Jørgensen, Rolf Egil and Gerd Ingrid Larsen, Reidar Maeland, Birgit Njåstad, Inge Marie Olsen, Tormod Sandaker, Kjell Skog, Kjetil Nielsen Skog, Svein Solberg, Helene Solvang, and Elisabeth Valmyr Åberg; at Alta, I thank Einar Eythórsson and Sveinung Eikeland; at Havnes, I thank Einar Giæver; and at Perth, Australia, where I spent the winter of 1997 transcribing interviews, as an Honorary Research Fellow at the Department of Anthropology, University of Western Australia, I thank Bob Tonkinson, Kate Morse, and Guy Wright.

My work could not have occurred without the patient support of the people of Arnøya, who welcomed us into their homes and hearts. Although ethnical protocols require that most remain anonymous, appearing with pseudonyms, I single out Rolf Egil Larsen and Ottar Friis, and thank them for their many contributions to the project, invariably offered with candour and a smile. Similarly, Svein Jentoft has been steadfast in his support, first suggesting Arnøya as an appropriate field setting and subsequently helping in a thousand different ways. Without Rolf, Ottar and Svein this book might never have been completed. Finally, I thank my wife, Karen Olsson and our sons Alexander and Erik, all members of the 1995 expedition. My family supported my determination to see this project through to completion. Most especially, Karen has been of tremendous assistance in formatting the manuscript, making maps, hearing me out, and just being there. In thanking all those listed, I accept complete responsibility for all errors of fact or interpretation.

Preface

Voices

An Årviksand youth on out-migration

Folks like me go to bigger towns. They go to the military and then they're gone. I lived two years in Skjervøy, going to school, and missed Årviksand. I missed the silence. I like the place very much but I don't want to work here forever.

An Akkarvik woman on out-migration

I think it's sad, but people have always moved. But on the other hand, it shouldn't be impossible to live here. You don't have jobs here but some commute. People [will] still move, even if you get the jobs and shops.

Haugnes fisherman on out-migration

When we close this [house down, either die or move] then it will be dark out here, and I wonder how long before the rest of Arnøya will take to follow

A Lauksletta man on centres and peripheries

People living on Skjervøy don't see the need for Arnøya, Laukøya. They need the small communities around the centre and a link between the small communities and the centre. Just like the wheels on a bicycle, to keep going with the bicycle you need both the centre and the wheel around, and the links between the wheel and the centre. They need the districts to keep the centre going. Namely, as you say, they produce services. Two people can't make a living out of cutting each others' hair. One can make a living out of cutting the other guy's hair but both cannot. So they need the third person, and that third person is living on Arnøya.

An Arnøyhamn man on local knowledge

When I was ten or eleven years old I started with my father to be in the boat. When I was 14, I knew all the fishing places. You see we take the landscape and angle the different places, I knew all the places from the landscape, outside, when I was 14. What I learned was absolutely necessary for me during all the time I have been in the boat, until today. You didn't have an echo locator.

An Arnøyhamn man on the fishing quota system

Before the crisis everyone could buy his own boat and start to fish. So it was easier to be a part of the fishery. The difference between the early 1990s and the last of the 1990s was that the quota system that the government [introduced] made it very difficult for new people, and especially young people, to have his own quota and fishing licence. That was something you didn't need before. So, it made it nearly impossible for new recruitment in the fishery.

A Tromsø fish plant owner on oil and fish

Norway [See Figure 1] *is special because of the huge amount of oil which is pumped out of the North Sea. This is destroying all land-based industries. Norway is putting too much of its eggs in this one basket.*

A former Årviksand resident on community integration

In the 1960s and 1970s, religion held the community together, in a common way of thinking, combined with the agriculture and fishing, forming a whole pattern.

An Arnøyhamn man on technology and social cohesion

It is interesting that since we have these opportunities to go shopping by internet, dealing financial stuff over the phone, with the mails and so on, the contact between people is lost. When you are going to the shop the cashier has little to do but punch in the numbers, and you leave, go home, sit down by the TV. You are losing contact with people. This hasn't occurred yet so much on Arnøya. My parents and grandmother are always talking about other people visiting you, all the time. Every afternoon you had someone visiting you. That is also gone. And we are losing contact with people. The only people you see are your own family or the people on TV. Maybe you eat breakfast with your family but then you go out to work, to hobbies, sports, and are more and more losing contact with other people. That scares me a little.

A Lauksletta man on capital flowing south

All the fish is going south. You have Oslo. They get the profit of export. Because they're driving [fish] from northern Norway, down to Oslo, and then into Sweden and Denmark. And that gives rise to the notion that wealth is produced there. But its only that is where the road goes through. The value is actually being produced here, and in Finnmark.

Opinions and polemics, hopes and fears, wisdom and common sense, a small sample of the voices of people from one region on the northern edge of Europe. I collected these and other narratives during anthropological fieldwork on the North Norwegian island of Arnøya[1] in 1995 and 1997 (see Figure 2). What follows is my account of the people of Arnøya, and of their island 'communities'[2] told in their voices, about their struggle to survive in our rapidly changing world. Although unique in many ways, Arnøya faces some of the same challenges as other rural regions of the world. I say more about Arnøya and the plan of the book below, but first attempt to place the Arnøy case within a global context.

Like their counterparts in the Third (underdeveloped) World, rural peoples living on the periphery of First (developed) World Nation-states[3] are often threatened by forces over which they have decreasing control. Rural economies dependent on extracting natural resources such as trees, agricultural products, minerals, or fish, are endangered by resource depletion, by rising energy prices, by fluctuating international markets and currencies, and by neoliberal trade agreements that favour free trade and other forms of laissez-faire economics, and discourage subsidies to peripheral regions.[4] First World farmers, including many in Canada, the United States, Europe, Australia, and elsewhere, face growing demands to modernize, so as to compete in the increasingly global market place.[5] In addition to these economic and political pressures, the globalisation of mass media and communications generally depict urban imagery, subtly depreciating the contemporary relevance of rural peoples and the places where they live, and conveying the message that better futures are to be found in urban centres. These

[1] The Norwegian place name for the island of Arnøya ends in the letter 'a' meaning 'the'; when the island's name is used as an adjective, such as 'Arnøy people' it lacks the letter 'a.' The prefix of Arnøya is derived from an archaic Norwegian word ardni, meaning 'eagle'; øy means island. Arnøya literally means 'eagle island.'

[2] Anthropologists continue to study 'communities,' despite disagreement over what the term means (c.f. e.g., Rapport and Overing 2000:60-65). Some consider the term too problematic (cf. e.g., Rapport, in Amit and Rapport [2002]) or use neologisms, such as 'locality,' to refer to 'communities.' In what follows, I use 'community' to refer to the parishes or villages of Arnøya. In doing so, I follow others, such as Paine (1965) and Thuen (1995), who refer to Nordbotn and 'Nordfjord,' respectively, as 'communities.'

[3] I acknowledge that rural peoples in developing or Third World countries face far greater challenges than those described here, primarily because social welfare programs are less available in the Third than in First world nation-states.

[4] Hovgaard (1998: 48) distinguishes between peripheries, where basic production (extraction) takes place, and centres, places of marketing and finished production. While this remains true, in future some 'niche-marketed' products, produced in rural areas, could be important additions to traditional industries, and contribute to rural viability.

[5] See, for example, Adams (2002) for threats to North American farms, and Gray and Lawrence (2001), for similar problems in Australia.

and other reasons fuel the rural exodus so obvious in many First World censuses,[6] and invite the question who, in the future, will live in rural regions to produce the foodstuffs and other raw materials necessary to live in tomorrow's growing centres? Yet those who argue that rural regions are simply a drain on the public purse may ignore this question.[7]

Contrary to these trends and beliefs, I believe that there are sound economic and strategic reasons to support policies and programs aimed at rural revitalization. Centres and peripheries have long been economically and socially interdependent, with people and goods flowing both ways. Rural depopulation threatens this interdependence and lessens the diversity of ways of living in contemporary nation-states. The exodus of rural areas might well be used to justify privatisation of 'common property,' and thus hasten the move from state to corporate management of the 'commons.'[8] Moreover, vacant hinterlands may open the door to corporate management of resources rural peoples have successfully managed for generations, opening the door to corporate management, with, I maintain, devastating consequences for natural resources. And finally, depopulated hinterlands may encourage economic or even military challenges to national sovereignty.

Of course, further depopulation of the rural regions of First World nations may justify corporations seeking new sources of raw materials in another, less developed part of the world. This latter trend, one meaning of the term globalisation,[9] connects the First and Third World in a new relationship, one which bypasses producers in First World peripheries in favour of those of the Third World, where raw materials and finished products are produced more cheaply, and where workers are usually not protected by the umbrella of social programs common to progressive nation-states, such as Norway or Canada. For me, globalisation is capital (commonly First World industrial wealth) seeking out and employing ever-cheaper labour, so as to increase profit. Used in this sense, globalisation is not new, although today's communications technology means that it is more rapid and potent than ever before. Although globalisation may be good

---

[6] The 2001 Statistics Canada census revealed increased urbanization and the growing urban-rural demographic imbalance.

[7] Canada's example of this is the Vancouver-based Fraser Institute, which berates the people of rural regions, and discounts the actual value of goods produced there.

[8] One example of this is the move from Individual Vessel Quotas (IQs, as discussed in Chapter 5) to Individual Transferrable Quotas (ITQs, as in Iceland), which transforms use rights to common property (and thus to fish) into private capital, owned in ever-greater amounts by large, often transnational, corporations. Another, similar example of privatization, that of 'trust agreements' in the fishery, is presented below, in Appendix 2.

[9] There are many definitions and opinions about globalization and its positive or negative implications for First World peripheries, such as Arnøya (See, for example, Rosenau [1997]; Weber [Ed,1998]; Saul [2005]; and others). Most definitions of globalization, as well as predictions about its consequences, identify new linkages between economies or societies formerly considered isolated from one another.

for corporate 'bottom lines,' its long-term economic and social costs may be less desirable for rural peoples in both the First and Third world.

There are also many ways of depicting the rural regions of contemporary First World countries, and even reasons to question the degree to which rural regions actually differ from urban centres, insofar as the peoples of both consume as well as produce. Yet we may know less about rural regions today than a generation ago. My sense is that fewer academic researchers now study and report on small-scale hinterland communities, than formerly. Curtin (1997: 256) makes precisely this point in relation to decreasing ethnographic research on rural Ireland, explaining how this diminishes our understanding of many contemporary phenomena, such as the role of 'newcomers' or 'strangers' in Irish villages. Possibly, the pursuit of theoretical 'relevance' makes it passé to conduct field studies in small-scale, rural communities, even though such research potentially has much to say about how the peoples of nations-states see themselves (Herzfeld 1997:180). Another obstacle to rural research may be post-modern academic theory, which often questions the very concepts (such as community, see for example, footnote 2, above) that in the past have been useful (albeit imperfect) starting points for studying the places where rural people live.

Well out of range of such scholarly debates, rural struggles continue, usually demonstrating the growing dominance of urban centres but sometimes revealing surprising rural triumphs. Notwithstanding the current fashion away from rural community studies, this study presents the results of field research in one First World periphery. My goal is to understand how the lives of Arnøy people and the places where they live have changed, especially since the Second World War. The actual voices of islanders are centre stage, although my intention is less personal than sociological, and my focus more on the *places* where people live, than on individual lives. Pseudonyms replace the real names of all living Arnøy people quoted in the text. We attempt to learn why the populations of some island villages have grown, or held their own, while others have declined? Has the agency of islanders helped their villages survive? My inquiry goes straight to the people themselves, and my task involves sifting through the information I collected (See Appendix 1, for methodology) in an attempt to reconstruct what happened. In presenting these results, I realize that some may find my account excessively detailed, historic, ethnographic, or even in places, simply wrong! [10]

What explains differential community viability? We shall see that the economic, social, and political factors responsible for community viability are both local and external and are continually changing. The position taken here is that one community's viability results from the complex and ever-changing interaction

---

[10] The fine multidisciplinary, international study by Apostle et al (1998) provides the contextual background for the more localized and grounded material presented below. On the issue of my research on Arnøya, one person warned me that the Norwegian '*Jente* law' disapproves of outsiders from writing local history.

between many factors, rather than any single cause. Under certain conditions, multiple factors may create what is in effect a chain of causation, enhancing or reducing the possibilities for community viability. Finally, when one considers our urbanizing world, I discuss the future of rural areas such as Arnøya. Here my appeal, directed at urban centres of power and capital, urges rural revitalization.

My interest in community viability emerged after I began fieldwork. Originally, I wanted to study Norway's fishery crisis so as to better understand the on-going ground fish crisis in Atlantic Canada. Thus, the impetus for my research in Northern Norway originated from events occurring in Newfoundland-Labrador, Canada, where I have conducted anthropological research since 1971.[11] Chief among these events was the closing of the five century-old northern cod fishery, in 1992.[12] A brief overview of Atlantic Canada's cod moratorium is presented in Appendix 2. Because North Norway had closed its arctic cod fishery, albeit briefly, in 1989, I wondered what might be learned by examining a similar region that had also lost its fishery?

Although the fishery remains important in what follows, early in my fieldwork it became apparent that the future of two of Arnøya's four small communities seemed particularly doubtful, for reasons that were at the time not clear. The issue of community survival appeared paramount to local people, and this prompted me to redirect the focus of my research. Although the fishery and regional fishery politics appeared crucial to community survival, my interviews suggested that many additional factors, ranging from the provision of community services, to lobbying by local political groups, to transportation policies, affected the differing fortunes of each small village. These observations, unknowable prior to fieldwork, caused me to alter my original focus to examine how the island's fishing communities cope, not only with the booms and busts of resource-dependent economies, but with regional and global forces.

The Chapter Format

In the chapters that follow, the people of Arnøya tell their own story, although I bear responsibility for the chapter topics and the narratives that appear in the book (a detailed account of my methodology, including my use of local voices, is provided in the Appendix 1). The first three chapters provide necessary background for understanding contemporary Arnøya. Chapter 1 provides a brief overview of Norway and its two forms of regional government -- provinces (*fylkes*) and municipalities (*kommuner*) and then introduces Arnøya. The remainder of Chapter 1 provides contextual information about the environment and history of the Arnøy region, such as the historic fisher-farmer (*fiskarbonde*) economy, which preceded

---

[11] My 1995 book is closest in focus and intent to this study (Kennedy 1995).

[12] See Cadigan (2001) for an excellent review of the growing literature on fishing moratoria in Atlantic Canada.

today's industrial fishing economy. Chapter 2 continues my chronologically ordered description of context, explaining, for example, the local significance of the Second World War, an event with real and lasting consequences for Arnøya. Chapter 3 hones in on important changes occurring on Arnøya during the two decades prior to the Norwegian fishery crisis of the late 1980s. Significantly, these two decades witnessed the end of the older pluralistic fisher-farmer way of life and the introduction of the industrial fishery. Chapter 4 discusses the fishery crisis, which officially began in 1989. I contend that the crisis was not confined to the oceans; the impact of the fishery crisis was intensified by a financial crisis ashore, causing economic restructuring that I trace to falling international oil prices. Chapter 5 describes the contemporary fishery, and begins with my journal entry from a 1995-fishing trip to the coast of Finnmark aboard what was then the largest fishing vessel on the island. Chapter 6 explores recent regional transportation policies, specifically the development of roads and ferries. These and other components of infrastructure are vital to the island's future (a point made in Chapter 8), but Chapter 6 examines another, consequence of transportation policies, namely how road and ferry connections favour some villages more than others, which in turn fuels periodic competition and consensus between the island's villages. Chapter 7 examines out-migration, mainly of individual youth, now a common chapter in most Arnøy biographies, as is the case in rural regions around the world. Here, and in the concluding chapter, I advance the major policy implication of the book, namely, that for rural regions to survive, they need to have infrastructures (schools, medical services, roads, and so on) roughly comparable to those of towns and cities, motivating some people who leave their home villages, to return. Chapter 8, the concluding chapter, briefly reviews the topics covered in earlier chapters, focussing on what contributes to community viability, a term I prefer to the contested yet more commonly used term, sustainability.[13] I conclude by restating my argument that improved rural infrastructure and return migration reduces the negative consequences of out-migration and offers hope for the future of rural regions such as Arnøya. And finally, an Epilogue based on my return to Arnøya in 2003 briefly updates the story.

---

[13] For me, the term sustainability suggests the *continuation* of some earlier way of life that, for many reasons, may no longer be possible. See Lelé (1991) for a good overview of sustainable community development, and Ommer (2002), for case studies further illustrating the potential for such development within the context of Newfoundland's fishery crisis. A related literature proposes terms such as 'retrenchment,' to refer to reduced fishing following restructuring (See Pettersen 2000: 85) and 'regeneration,' to refer to conservation by people awaiting better times (See Byron 2003). Rather than any of these terms, I prefer the term 'viability' when applied to a restructured society and economy that 'works' but may (or may not) be directly related to previous socioeconomic forms (See Saugestad 1988 for a focus on 'household viability').

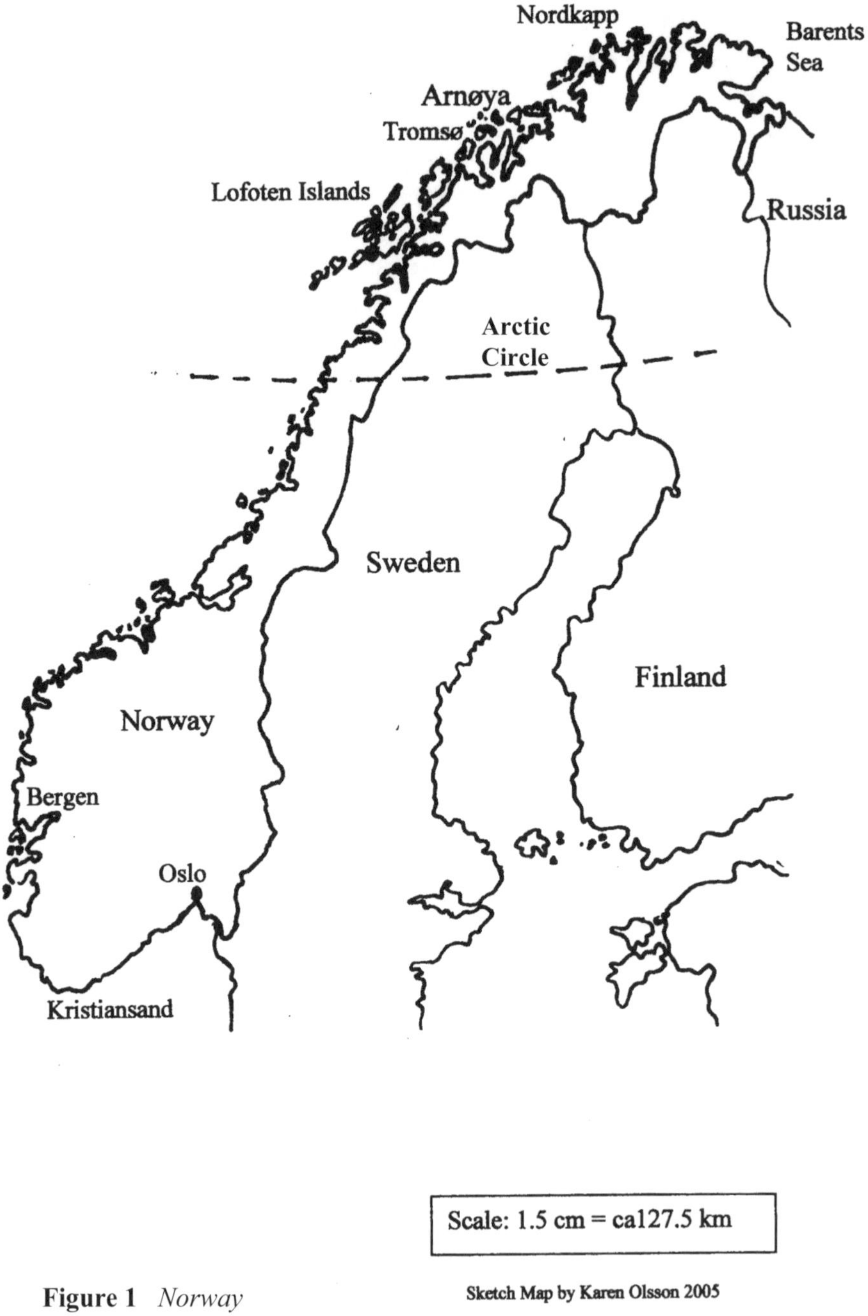

Figure 1 *Norway*

CHAPTER 1

HISTORICAL CONTEXT

This and the next chapter presents the historical context helpful for understanding the contemporary narratives presented in this book.

I first briefly describe Norway, its counties, and municipalities. I then outline Arnøy's natural environment, the island's prehistory, settlement by Norwegians, and adaptations up to and including the 19th century. Given my focus on settlement concentration during more recent decades and latterly, on community viability, Arnøy's historically decentralized settlement pattern and distinctive mixed fishing-farming (*fiskarbonde*) economy contrasts rather sharply with today's more centralized settlement pattern, based as it is on an industrial fishery. The intention of this and the next chapter is to outline the essential 'before' for data presented in chapters 3-7.

Country, County, and Kommune

The four and one half million people of Norway live in a long, thin country, bordered along its west (or south) coast by the Atlantic Ocean and further north, by the Arctic Ocean (see Figure 1). Norwegians say that if Norway was pivoted at its base, say from Kristiansand in the south, that the northern tip of Norway at Nordkapp (North Cape) would reach southern Italy, or as Wikan puts it, a distance of from Maine to Florida (2002:31).

Although an active member in the world community, Norway twice (in 1972 and 1994) rejected membership in the European Union (EU), partially because of concerns about its fisheries. In 1992 (effective in 1994), Norway and other members of the European Free Trade Agreement (EFTA) joined the European Economic Area (EEA), enabling participation in EU programs and actions. Participation in the EEA provides Norway access to the larger European market, and in principle, decisions about fisheries and agriculture are negotiated outside these agreements. However, continued growth of the EU and European interest in Norwegian fisheries portends further developments on this issue.[1]

There are many exceptions to the stereotype of a homogeneous, rural Norway. Norwegian census data confirm that most Norwegians live in rural areas, even though the capital city of Oslo has over one-half million people and the second largest city, Bergen, has over a quarter million people. Norway's ethnic composition is relatively homogeneous, even though Oslo contains 158 nationalities (Wikan 2002:32). These nationalities, along with the indigenous Sami

[1] This paragraph is based on *http://europa.eu.int/comm/external_relations/norway/intro/* and on Apostle et al (1998: 100-102).

and historic Finnish immigrants (*Kvens*), identify themselves as distinct from the majority population. Norwegians speak Norsk although the language has two variants: Nynorsk and Bokmål, both taught in the school system, and each with its advocates and opponents. And as might be expected, there are also regional dialects, and indeed, even inter-community dialectical differences, such as exist on Arnøya. For example, my local research assistant required the spouse of a person from elsewhere in Norway, to translate.

The oil-driven Norwegian economy underwrites a generous social welfare system and standard of living with few international equals, even though there is both an emerging ethnic underclass, marginal and stigmatized, although not necessarily materially poor (Wikan 2002: 59-61; 242) and cases of poverty in many communities, including those on Arnøya.

Norway is divided into 19 counties (*fylker*) or provinces, similar to yet more decentralized from the national government than Canadian provinces or American states. Northern Norway includes the three northern counties of Nordland, Troms, and Finnmark. Each county includes a number of municipal governments or kommuner, such as, in this case, Skjervøy kommune, of which Arnøya is an outlying part. Let us move now to Arnøya.

Arnøya,[2] site of this case study, is one of several outlying islands in Skjervøy municipality (*kommune*), in northern Troms County (*fylke*). The mountainous and ruggedly beautiful Arnøy landscape comprises approximately 270 square kilometres, while that of Skjervøy Island, location of the *kommune* capital, is only 11 square kilometres (Helland 1899: 321). Arnøya is accessible only by ferry while Skjervøy is now connected to the mainland by a bridge and a tunnel (see Chapter 6).

There were four parishes or communities (*bygder*) on Arnøya (See Figure 2) and in 1995 and the island's population was 509. One travels to the island by car ferry, and docks at Skardet, within the community of Lauksletta. Lauksletta (and the neighbouring island Laukøya) is named after a type of local onion or chive (*lauk* or *Allium sibiricum*) introduced by historic Pomor traders from Russia. Lauksletta is dispersed along Lauksund, the sheltered strait separating Arnøya from Laukøya. In 1995, Lauksletta had a fish plant, salmon farm, and general store; yet that year its thirty-five year old school closed, and its four remaining students began attending Arnøyhamn School[3]. Proceeding west along Route 347, the two-lane paved road snaking, as Norwegian roads often must, along the seashore, one passes below Singeltinden, a 678 m mountain feared for winter avalanches (See Figure 2, roads are in broad black).

---

[2] Geographically, the Arnøya discussed here might be considered 'north' Arnøya, as there is another Arnøya just south of Bodø.

[3] A local teacher who began her career at the Lauksletta School in 1969 recalled that there were 56 students her first year and only four grade three students in 1995, the year the school closed.

The next community, Arnøyhamn, runs along the road above the shallow waters of Grunnfjord. Particularly since the Second World War, Arnøyhamn has become Arnøy's administrative centre, boasting an elementary school, the island's only junior high school (grades 8-10), a small general store, a weekly medical clinic, a senior citizens apartment building, the island's only Lutheran church and cemetery, a building supply outlet, an underdeveloped industrial zone, and several other small businesses. Continuing north around Langfjord one arrives at the intersection of Routes 347 and 348, the latter road leading to Akkarvik.

Akkarvik has the best natural harbour on Arnøya. The community stretches several kilometres around Akkarfjord and is overlooked by 850 m high Trolltind Mountain. In 1965, construction of a radio transmitting station began on Trolltind. Akkarvik was once Arnøy's largest community but is today small and dying.

Rejoining Route 347 and proceeding north, one reaches Årviksand, Arnøy's largest community in 1995. Årviksand's stone breakwater *(molo)* is both the community's foundation and its main weakness. The breakwater enables Årviksand to have Arnøy's largest fish plant yet requires periodic dredging and repair, restricting the size of vessels supplying the fish plant, and thereby threatening the viability of the community. Having introduced Arnøya and the broader nation-state of which it is a part, we turn now to an overview of the natural environment and history of the island.

The Natural Environment

Arnøy's exposed location and dependence on the fishery exemplifies north Norway's maritime ecotype.[4] Much of the island's northern coastline lies exposed to the Norwegian Sea, while its southern coastline is more fjordal, protected by islands adjacent to the Norwegian mainland. Like much of the northern Norwegian coast, Arnøya is mountainous, with narrow and well drained coastal plains bordering rivers at places such as Nord Rekvik, Lauksund, Grunnfjord, Akkarvik, and Årviksand, places which are not surprisingly sites of past or present human habitation. The interior of Arnøya contains some small lakes, such as Skartindvatnet (0.4 km square) and Langfjordvatnet (0.4 km square).

On higher elevations, one encounters lichens and other alpine vegetation, as well as rock faces (*hammeren*), often too steep to climb. Locals characterize the grasses (*eng*) comprising Arnøy pastures as rich, containing, they contend, some of the best fodder for grazing animals (resident domesticated sheep; visiting Sami reindeer) in northern Norway.[5] At lower elevations are bogs (*myr*), formerly mined

[4] Paine (1965: 13-17) contrasts North Norway's maritime pattern of exploitation, where the focus of exploitation concentrates on one resource, such as fish, with the fjordal ecological pattern, where exploitative activities are divided seasonally between several resources.

[5] A less complimentary evaluation of Arnøy pastures occurs in an historical book of topographical statistics, published in 1899, which rates the island's pastures as poor, supporting only 800 reindeer

Sami Reindeer admist bjørk, Langfjord 1995. (Photo by Karen Olsson)

by locals for peat, once the dominant heating and cooking fuel. Several species of berried plants grow on such bogs, including the prized cloudberry (*moltebær*). Coniferous trees are absent on the island although there are some small birch trees (*bjørk*) in protected areas, such as around 8 km long Langfjord and Grunnfjord, and along rivers and on coastal plains. Rodents, such as the lemming (*lemen*), as well as introduced rabbits (*hare*), sustain larger predators including weasels (*røyskatt*). Otters (*ote*r) are occasionally encountered along rivers or in harbours. Neither moose (*elg*) nor reindeer (*rein*) are year-round residents, although moose occasionally swim to the island from Kågen (as two did, for example, in 1995), and each summer the Kautokeino Sami herd their reindeer there.

Several species of birds common to the land include the hooded crow (*kråke*), magpie (*skjære*), raven (*ravn*), and smaller birds like bullfinches (*dompap*), snow bunting (*tornrisk*), and house sparrow (*gråspruv*). As in northern Canada, two species of ptarmigan, mountain or rock (*fjell rype*), and willow or lowland (*li rype*) ptarmigan, and the spruce grouse (*skog rype*) live on Arnøya. Arnøya has one of the largest concentrations of eagles in northern Europe; three species, the white-tailed

---

each summer (Helland 1899: 328). Each summer, Sami reindeer feed at several pastures, including Lovlavarre, Givavagge, Rekvikdalen and others (Guttormsen 1993a). Current estimates of the summer reindeer herd are as high as 2500 animals.

eagle (*havørn*), the golden eagle (*kongeørn*), and the smaller, fish eagle (*fiskørn*)[6] are native, and seen quite often.

Sea birds include the common (and protected) eider (*ærfugl*), and mallard duck (*stokkand*); the great black-backed (*svartbak*) and herring (*gråmåke*) gull; the kittiwake (*krykkje*); puffin (*lunde*); and others.

Sea mammals include the porpoise (*nise*), minke whale (*vågehval*), the Greenland seal (*Grønland sel*), and others. Ocean fishes important to the local economy include capelin (*lodde*), herring (*sild*), redfish (*uer*), lumpfish (*rognkjekse*), cod (*torsk*), saithe or pollack (*sei*), Greenland halibut (*blåkveite*), salmon (*laks*), and others.

Cod is central. People distinguish between three varieties or 'races' of cod, even though opinions differ about these cod populations, agreeing only on *skrei* and *tare torsk*. Thus, local people distinguish between two kinds of stationary cod stocks: bay cod (*tare torsk*), discernible by its brown-green colour (reflecting its fjordal diet), and coastal cod (fjord *torsk*). A third race, Norwegian Arctic cod (*skrei*) travels an annual circuit between in the Barents Sea and the Lofoten Islands.[7] Each winter, six-year old spawning cod from Russia's Barents Sea swim south to Norway's Lofoten Islands (south of Arnøya) to spawn. In spring, the warm Gulf Stream carries cod spawn north to the Barents Sea nursery. There, the young cod develop and the cycle repeats.[8] Historically, but to a lesser extent today, dense concentrations of spawning cod attracted fishers from all along Norway's west and north coasts to the Lofoten winter fishery (*Lofotfisket*). An older Akkarvik man described details of local knowledge concerning varieties of cod, beginning with those fished in the Lofoten fishery: ***Skrei fiske*** *[refers to] about 4 kilo cod of the Lofoten fishery.* ***Skrei*** *can range up to 30 kilo.* ***Skrei*** *are a "separate race".* ***Skrei*** *are breeding cod, both male and females. They only go to Lofotens to breed, they swim outside, never come in here, never.* ***Skrei*** *migrate between Svalbard and [the] Smutthullet and the Lofotens; they are not the same as fjord* ***torsk****, not the same.* ***Tare torsk****, fjord cod are smaller, more brown.* ***Tare torsk*** *doesn't go out in the ocean. Finnmark* ***torsk*** *stay in Finnmark. [So] there are at least three kinds of* ***torsk****.*

---

[6] Locals consider the *fiskørn* an eagle. Scientifically, it is an Osprey (*Pandion haliaetus*) and in a different genus than eagles, such as the white-tailed eagle (*Haliaeetus albicilla*).

[7] Based on interviews with 70 fishers from Finnmark, Maurstad and Sundet (1998) describe three cod stocks north of 62° North Latitude. Two of these three correspond with the last two races of cod mentioned above, while a third, 'Murman' cod, are fished off the Russian coast. Fishers reluctantly told the two researchers of 44 local spawning grounds of fjordal cod, but worried that divulging such valuable local knowledge could again lead to over fishing by non-resident fishers.

[8] Since 1976 Norway and Russia have jointly managed Barents Sea cod, claiming 90% of the total catch and allocating the remainder to third parties, such as the Faroese (Hoel 1991).

Russegrave, with Årviksand in background. (Photo by Karen Olsson)

Precipitation in rain, fog, and snow are common to Arnøya, even though when humidity is low, the air is surprisingly dry for a maritime climate. Snow covers the ground to depths of one to two metres for roughly half the year, but melts rapidly in the warmer temperatures and increased daylight of May and June. Light frosts are common between January and March but summer can be remarkably balmy. Winds, such as the prevalent land wind (*landvind*) are strong throughout much of the year. The seas surrounding Arnøya affect all facets of human life. In Eastern Canada, the Labrador Current creates a sub-arctic climate at the same north latitude as the United Kingdom south to France. Whereas in northern Norway, the North Atlantic Drift of the Gulf Stream warms the oceans enabling life forms and economies inconceivable in comparable northern latitudes.[9] For example, in late May 1995, I accompanied an Akkarvik salmon farmer to his salmon cages where

[9] Writing about 'Fairisle', just to the south of Arnøya, Saugestad (1988, 51ff), observes that when compared with other northern regions, the Norwegian coastal environment at 70 degrees north latitude is anomalous.

ocean temperature readings measured 8.5° C; similar readings for comparative depths and dates in Newfoundland may only reach 1° C. The consequences of the warm Gulf Stream for life in Northern Norway should not be underestimated and need to be centre stage in any comparisons with Atlantic Canada.

Prehistory and Early History

Much of Scandinavia south to northern Germany remained ice covered during the last glaciations around 13,000-14,000 years before present (B.P.). The earliest cultural remains on Arnøya are small Stone Age house remains at Elva, on Kågsund (Holberg 2002: 39-43). Next and nearer to Arnøya is the multilayered site at Skattøra, overlooking the sound between the islands of Skjervøy and Kågen, and dating from 6,000 B.P. Late Stone Age (4500-1800 B.P.) hunters at Skattøra harpooned seals passing through the sound.[10]

Knowledge about the past increases as we move forward in time to the Iron Age, the first millennium AD, although whether Arnøya was occupied during all of the Iron Age, and if so, by whom, awaits further archaeological investigation. At issue here is whether Arnøya was the northern edge of the southern-based Iron Age, and, as appears more certain, that Arnøya was within the boundaries of a more northerly situated *Russegrave* complex.[11] The existence of numerous *Russegrave* sites on Arnøya confirms that the island was near the southern limit of *Russegrave* complex, dating from the Iron Age; the *Russegrave* complex is not found south of Karlsøy *kommune*.[12] Roughly half of the 400 *Russegraves* from North Troms to eastern Finnmark are located on Arnøya and nearby Laukøya, typically close to the seashore. *Russegraves*, coffin-shaped and sized (3 by 1.5 m) depressions once mistaken for human graves, were once lined with flat rocks and sea mammal skins, and apparently used like cauldrons to render seal and whale blubber into oil. They may have been used by prehistoric Sami, to produce oil owed to chiefly families.

---

[10] Johnson (1996: 18-19). An excellent inventory of prehistoric and historic cultural remains from Arnøya/Laukøya is found in Holberg (2002). Holberg's research occurred under the Bygde-mobiliseringprosjektet (community mobilization project), successor to the Kystsamfunn-Prosjektet, discussed in Chapter 7.

[11] The early history of Arnøya during the Iron Age is unclear. A 9th century chieftain named Ottar claimed to have sailed three days northward (presumably from his base at Kvaløy [Tromsø]) along an uninhabited coast (Lund 1984, quoted in Perdikaris 1999), implying that Norwegian and/or Teutonic speaking peoples settled Arnøya after that time. Similarly, historian and archaeologist Håvard Dahl Bratrein's (personal communication, 1995) research on the Helgøy region, just south of Arnøya, investigated whether Iron Age occupations existed further north of that region. Bratrein is uncertain, but adds that several Iron Age fishing or whaling sites at Loppa (just north of Arnøya) may mean that Arnøya falls within the northern boundaries of the Iron Age.

[12] Details of the excavation of a Russegrave site at Lauksletta are presented in the Årbok for Nord Troms(1990). See also Guttormsen (n.d.) and Holberg (2002: 23-25).

One of the only *Russegrave* on Arnøya that has been excavated (by archaeologist Kjersti Schanche, in 1990) produced charcoal dated to between 300 and 800 AD. Although North Norway has long been rich in marine resources, since the Iron Age it has suffered from a chronic shortage of capital, requiring an influx of credit from various sources.[13] Like peasants elsewhere, control of the riches of the land and sea, fish, furs, seabird colonies, pastures, and so on, rested with absentee owners (first Norse or Viking aristocrats, then the King or Catholic Church, and finally, southern Norwegian capital), meaning that until the early 20th century (and some might say still), value was extracted from North Norwegian producers to benefit economic interests elsewhere. Although proof of when Norwegian and/or Teutonic-speaking farmers/fishers settled Arnøya awaits further archaeological research, research findings from the Helgøy region to the south may apply to life on Arnøya as well.[14]

Sami and Norse settled the Helgøy region during the post-Viking era, each adapting to separate ecological niches, and peacefully co-existing. The Sami tended to live in more protected 'inside' coasts, while the Norse lived on exposed 'outside' coasts. Residential separation may be explained by one of two reasons. The first is economic, reflecting the Norse concentration on fishing and the Sami hunting of wild reindeer. A second explanation is political, and linked to taxation. At the time (the Middle Ages), the Russians, Danes, and Swedes taxed the Sami, while only the Danes taxed the Norse. Consequently, the residential separation of Sami and Norse may have been a policy of the Russians and/or Swedes to prevent the Sami from mingling with the Norse, and thereby avoiding paying taxes.[15]

In the second half of the 16th century, taxation and Norwegian immigration from the south upset the pacific co-existence between Norse and Sami, a change that would eventually dispossess Sami from much their land base, leaving them more marginal.[16]

Between the 1300s and the economic depression of 1620, much of the Norwegian coastal population participated in trade organized through the Hanseatic

---

[13] My argument here follows Bratrein (1992) and Wolf (1966), on peasantry.

[14] The dearth of archaeological work in north Norway, work which may eventually clarify details discussed here, is clear from Holm-Olsen's (1981, cited in Perdikaris 1999: 392) observation that only 25-30 of the approximately 2000 farm mounds in north Norway have been excavated! Of course some areas, such at Helgøy, just south of Arnøya, are better known than others. An interdisciplinary project based at Tromsø University began in the late 1970s, investigating economic and social conditions in the region between AD 1250-1650, and their conclusions may apply to Arnøya during the same period. For some of the results of the Helgøy project, see Mathiesen (1990; 1993) and Bratrein (1992).

[15] Olsen (2000), following the earlier lead from Odner (e.g., 1985) argues, contrary to earlier notions that Norse settlers suppressed the Sami, that in fact the two peoples had a symbiotic relationship, which (following Barth, 1969) solidified Sami ethnic identity.

[16] My argument here draws on Thuen (1995: 25-26).

League, exporting stockfish and importing cereal grains.[17] Although much the land and resources in what is now Troms county was owned either by the local nobility, the Catholic Church, or other absentee owners, Bratrein (personal communication 1995), who has studied documents from the period from 1550-1600, believes that Sami continued to control Arnøya until around 1600, renting land to Norwegian newcomers. Bratrein's history of Karlsøy and Helgøy contains the 'Nordlandia' map dating from between 1740-50, depicting 'Arnoe' and both Langfjord and Grunnfjord (1990: 240).

Just south of Arnøya, Aubert et al (1970) describe absentee ownership of 'North Island,' during the 16th and 17th centuries. Until 1660, parts of 'North Island' were first owned by Danish-Norwegian crown, then sold to a man from Holstein (within what is now Germany), and then owned by private proprietors (*proprietærer*) until the 1840s. Thus, until the mid-19th century, the fisher-farmers of northern Norway, such as those of 'North Island' and likely also Arnøya, were tenant farmers, leasing land, sometimes from absentee landlords residing hundreds of kilometres further south.[18] Hovland et al (1982) argue that between 1750-1850, North Norwegian fisher-farmers participated in a proto-industrial fish exporting economy, much as did their counterparts in continental Europe prior to industrialization.

The earliest census that I am aware of that includes Arnøya is from 1801. It lists 27 people at Akkerfjord; 22 at Langfjord; 26 at 'Haunes' (Haugnes); 5 at Stagenes; 45 at Aarvig; 15 at Grunnfjord, 9 at Grundjiord; 7 at Løgstotten; and 10 at Singlen, a total population of 166 (See Historisk Institutt 1801). Excepting Aarviq (Årviksand), most of these settlements are 'inside' and face the mainland, leading me to suspect that most of these people would have been coast Sami (*Sjø Same*).

Throughout North Norway the mid-19th century witnessed the beginning of a major cultural clash between Norwegians and Sami. Between 1850-1960 the Norwegian state administered an assimilation policy of 'Norwegianization', (*fornorsking*) promoting everything Norwegian and dismissing most Sami customs and beliefs as backward.[19] Coincidentally, the Norwegianization of Sami occurred during a wave of Finnish (*Kven*) immigration to Northern Norway, between 1830-80, which increased the population of some areas. Norwegianization was rooted more broadly in 19th century Norwegian (indeed European) movements towards

[17] Historical details surrounding how north Norway was drawn into the Hanseatic stockfish trade are presented by Bratrein (1992: 40-41), Stensland (1990: 139), and Perdikaris (1999), among others. Perdikaris're-examination of zooarchaeological evidence from the Lofoten/Andøya area argues that stockfish production preceded the Hanseatic League, and functioned locally to sustain chiefly power.

[18] This paragraph draws on the comprehensive but sadly unpublished manuscript by Aubert et al (1970).

[19] Kvaal (2001: 196) notes that Sami children were required to speak Norwegian in school until 1970.

nationalism and modernization. Evidence of Norwegianization may be found in education, health and agricultural policies implemented by the Norwegian government, policies which considered the culture of Sami and *Kven* anachronistic and irrelevant to a modernizing national agenda. The Sami and *Kven*s responded to these assimilationist invectives by adopting Læstadianism (discussed later on), a religious sect within the state Lutheran church that espoused local rather than national goals. Sami and *Kvens* rejected Norwegianization by demanding that clergy reshape Lutheran rules to reflect local beliefs about appropriate economic strategies, dress, language, and so on.[20]

Until 1905 the mercantile town of Havnnes, in Lygnenfjord, had been the largest centre in the region. On the other hand, today's kommune capital, Skjervøy, was smaller and settlement was discouraged by both the Norwegian church and a resident merchant (*nessekonge)* named Nicolay Rasch. Thus, prior to 1905, only about 50-100 people lived at Skjervøy. Friis (1861) map shows a church in what appears to be seven Norwegian and/or Norwegian-Sami households. Thirty years later, Friis (1890) second map again shows the church, but few more houses. However, beginning in 1905, a new priest named Barmann began encouraging settlement, leading Skjervøy to grow larger than Havnnes.

Nineteenth Century Arnøya: Seasonality and Inheritance

Life on Arnøya during the nineteenth century bore little resemblance to life there today, and a brief examination of the past places the present in sharper focus. This examination focuses on themes relevant here, including settlement pattern, inheritance, and the seasonally ordered fisher/farmer economy.[21]

Farm homesteads were named after physical or cultural characteristics unique to where they were situated. Each farm included a number of out buildings used seasonally for specific purposes. Fugelsøy's (1974, 367) photograph of the Mellemjord farm on the nearby island of Vorterøy, illustrates buildings that were perhaps 'typical' of many historic farms. Farm buildings faced toward the beach or strand (*fjære*), and were well spaced. Fields were named; with outfields collectively called *utslåtte*r. Fugelsøy's photo depicts the summer barn for cattle (*sommerfjøs*), the main cattle barn (*fjøs*), dwelling house (*våningshus*), meat-hanging house (*stabbur*), boat/net house (*naust)*, and general storehouse (*lagerskur*). We can assume that farmsteads on nearby Arnøya during the same time period contained similar farm buildings.

Indeed, another one of Fugelsøy's examples comes from 19th century Arnøya, the farm of newcomer Bendiks Bæverdal, and contains buildings similar to those of

---

[20] Details on the clash between Norwegians and Sami, the state's policy of 'Norwegianization,' and Læstadianism come from Paine (1965), Bjørklund (1988; 1992; 2000), Gorter (1975), Thuen (1995) and Minde (2003).

[21] Details on 19th century Arnøya come from Fugelsøy (1974).

the Mellemjord farm. Bæverdal, a teacher who settled at Grunnfjord ('shallow bay,' now Arnøyhamn) in the mid-19th century, operated a farm with even more specialized economic activities than in the case of the Mellemjord farm. Fugelsøy writes that, *Bendiks Bæverdal took over the farm in 1866. He was a teacher and built the school, and may have also built the house; they called the midwife's house (**jordmorstua**). On the farm there were many small outbuildings, one for milk (**melkebu**), one for cheese (**ostebu**), one for a special cake (**sirupskakebu**), and there was also a baking house.*[22] In short, both the Mellemjord and Bæverdal farms contained various buildings required for the very seasonal and diversified way of life of the time.

Once land became freehold, in 1844, farmland that had previously been leased could be purchased.[23] Norwegian inheritance law (*Odelsloven*) stipulated that the oldest child (sons only, until 1975) had the right to inherit or purchase the whole farm. Like impartible inheritance elsewhere, one child (normally a son) acquired the whole estate, requiring non-inheriting siblings move elsewhere to live and to work, albeit with fewer obligations to their aging parents. The *odel* also stipulated that the son (and occasionally daughter) acquiring the farm, or in some cases the person such as a son-in-law purchasing it, had specific obligations to parents (or affines [in-laws]) who remained on the farm until death. Under the principle of *kår*, the son provided support to his parents who remained on the farm, while another principle, *kårrets*, referred to the five-year duration of such obligations. As Saugestad (1988) explains, good agricultural land has always been scarce in northern Norway and the custom of impartible inheritance suggests that the division of holdings damaged their viability.

The case of Bendiks Bæverdal illustrates how inheritance actually worked.[24] In 1892 Bendiks sold his farm to his son-in-law Abraham Mikalsen, who married Bendiks Bæverdal's oldest daughter, Manna Oline (b. 1857); Abraham Mikalsen took over the farm in 1892. The property's deed records that Bæverdal sold his farm to his son-in-law (Abraham Mikalsen), for 90 NOK, providing that he (Bendiks Bæverdal) and his wife should be taken care of for the rest of their lives. Abraham Mikalsen also agreed to prepare food and care for the three sheep that belonged to Bæverdal, to take care of the land, to grow potatoes for his aging in-laws, and to remove one half the yearly harvest of peat (*torv*, Bæverdal was responsible for digging his part). However, Bæverdal should dig his own potatoes in the fall, and if Bæverdal died first, Abraham must take care of Oline, the person who ultimately inherited all the 'goods' of the house.

---

[22] Fugelsøy (1974: 374), TS translation.

[23] Details on Norwegian inheritance law come from Saugestad (1988: 57-58); for impartible inheritance, see the classic studies by Homans (1975) and Arensberg (1968).

[24] See Fugelsøy (1974: 374).

In addition to a tendency toward patrilocality, my data reveal greater population mobility than one might have expected in such a 'traditional' and seasonally-ordered society: men and women moved to and from historic Arnøya.[25] Although some nucleated, densely concentrated settlements like present day Årviksand existed in 19th century Arnøya, fisher-farming homesteads were usually interspersed along the coastline of fjords, accessible to fresh water, peat land, level land for fields, places to haul boats out of the ocean or to moor them, and so on.

Maps, censuses, and other documents from the mid-19th century confirm that Arnøya was home to North Norway's three founding peoples, the Sami, Norwegians, and Kvens, as introduced above. For example, J.A. Friis' (1861) map shows 37 households at 14 named places around Arnøy's coastline. The pattern of where Norwegians and Sami tended to reside is consistent with the Helgøy project findings summarized above: namely, Norwegians tended to live in outer coastal settlements like Årvik and North Rekvik, and Sami (or part-Sami) preferred more sheltered inner locations like Grunnfjord (Arnøyhamn) and Elven. The ethnic composition of many households was complex, often including one or more persons who spoke Finnish or Sami. The cultural and linguistic diversity of households, especially households where the mother spoke either Sami or Finnish, is also recorded in the nominal census from 1865.

Following the 1865 census, beginning at Elven ('the river') and proceeding clockwise around Arnøya, we count a total of 253 persons, living in fifteen settlements. Table 1 lists the number of people living at each settlement. For example, in 1865, sixteen people lived at Elven. Since many of these abandoned settlements now fall within the bounds of Arnøy's four contemporary communities, I have included the contemporary communities in the centre column. Finally, the settlements listed in Table 1 appear at the same locations as place names listed on Friis' 1861 map.

---

[25] In order to better understand the history of settlement, post-marital residence, and (by implication) inheritance, I asked each of the 45 adults completing my questionnaire where their ancestors lived when they were children. As expected, their responses show a tendency toward patrilocality; generally men remain and women marry in. Thus, only 49 per cent of respondents' mothers lived on Arnøya as girls, compared to 69 per cent of their fathers. Percentages drop as we move back one generation in time. Thus, only 26 percent of their mother's mothers and 46 percent of their father's fathers grew up on the island. More surprising than patrilocality was the relatively high proportion of ancestors not native to Arnøya or to other communities with Skjervøy kommune. Thus, 40 percent of respondent's mothers and 26 percent of fathers did not grow upon on Arnøya or other communities within Skjervøy kommune. This high proportion of ancestors not raised locally increases as we move back in time. Fifty-five percent of respondents mother's mothers, 42 percent of mother's fathers, 44 percent of father's mothers, and 40 percent of father's fathers were not raised on Arnøya or other communities within Skjervøy kommune. These percentages demonstrate considerable mobility to and from the island during the late 19th and early 20th centuries. Apparently, for the viability of the older fisher-farmer adaptation to remain relatively persistent overtime, especially in conjunction with impartible inheritance, some of those born to local families had to leave.

Table 1 *Arnøy Settlement Census, 1865*

Settlement	Contemporary Community	Population
Elven	Arnøyhamn	16
Grundfjord	Arnøyhamn	18
Haugnes	Arnøyhamn	32
Langfjord	Arnøyhamn	21
Akkarfjord	Akkarvik	17
Lillevik	Akkarvik	2
Gjetvik		7
Sotnes		11
Rekvik Sondre		8
Pankekil		12
Aarvik	Årviksand	35
Rekvik Nordre		24
Nygaard	Lauksletta	33
Tyvdal	Lauksletta	12
Singla		5
	TOTAL	**253**

Source: Registreringssentralen for Historiske Data *(RHD), Folketellinga 1865 for Skjervøy*

A decade later, the 1875 census lists one fewer settlement (14) yet a larger population (365) than ten years earlier (RHD 1875). By 1875, localities now lumped within Lauksletta contained three relatively contiguous places (Nygaard, Loklet-Nygaard and Kjyvdal-Nygaard) with a total population of 79. This represents an increase from the 45 people inhabiting what is now Lauksletta a decade earlier, an increase explained by immigration of Finns (*Kvens*) around the mid-19th century.[26] Indeed, I suspect that the influx of *Kvensk* newcomers can explain much of the population increase on Arnøya between the 1865 and 1875 census.

---

[26] See Aas (1998).

A brief look at two *Kvenske* newcomers, Johan Johannesen and Johannes Mathiesen, suggests that late 19th century Arnøya was a more heterogenous society than it was at the time of my research.

Johan Johannesen, was born [of Finnish parents] in Sweden in 1828, and is listed in the 1865 and 1875 censuses. Ivar Bjørklund (personal communication 1995), who helped me understand these censuses, surmised that Johannesen *was among the first generation of* ***Kven*** *migrants. He's a Kvennish culture hero who [likely] walked from Northern Sweden to Arnøya.* The 1865 census lists the 38 year old and apparently widowed Johannesen as living with his three year old daughter, his adult sister, and the Swedish-born [of Finnish parents] wife (Else Brita) of a Sami man named Lasse Andersen [Lasse, Else Brita and their children appear in their own home in the 1875 census]. In 1865, Johannesen was a self-employed farmer-fisher, owning four cows, ten sheep, and two pigs. By 1875, Johan had married Inger (whose father was of mixed [*blandet*] ancestry and mother was Sami) and the couple had three children. More remarkably, by then Johan's 'household' totalled 19 people, several of whom were Finnish immigrants (in one case perhaps a relative of Johannesen) who likely heard back in then impoverished Finland about the possibilities of working for his 'rich' countryman in the Arnøy fishery.

In the second case, Johannes Mathisen (or Mattisen; whom locals recalled was once known as 'Lauksletta Johan'), is listed in the 1875 census as born in 1839 in Kvænangen Fjord. He was either a Kven or mixed Kven-Sami; his father spoke Finnish, his mother Lappish. Johannes' fortuitous marriage to the second oldest (of five) daughter (Karen Marie Larsen, b. 1842) of a well-off Nygaard family lacking sons substantially enhanced Johannes' prospects. Johannes and Karen Marie, and their two young sons, appear in the 1875 census, along with seven unrelated workers, all of Finnish or Sami descent. Consequently, seven of the eleven people listed with Johannes Mathisen's household in 1875 were servants, young men and women working in various capacities for this ascending farmer-fisher household head. In Mathisen's case, two of the seven unrelated people in the household were servant boys (*tjænestegutter*), two servant girls (*tjænestepiger*), and two others were fishers.[27] It was these largely young and landless workers, often Finnish, Sami, or of mixed extraction, which swelled most households and in turn, explain the island's population increase between 1865 and 1875.

I interviewed the grandson of David Eriksen, a Finnish immigrant forced from Finland by hunger, who arrived on Arnøya in 1866 to work on the farm and fishery of 'Lauksletta Johan.' My informant recalled that Johannesen eventually owned much of the northern end of Lauksletta, and was also referred to as Nygård Kongen, that is, the 'king' of Nygård. My informant also confirmed that his father, and

[27] Other work statuses used at the time for unrelated household occupants include female cook (*kokkepige*); babysitter (*barnepige*); housekeeper (*husholderske*); and others. See Brox (1965: 85-86), for a description of servant integration into the households for which they worked.

presumably others of second generation Finnish and/or Sami ancestry, could speak some Finnish and/or Sami.

By 1890, the date of Friis's (1890) second map, Arnøya had some twelve occupied and named settlements with a total of 49 households. Of interest is that Nygård contained the largest concentration of households (11), whose occupants were of mixed Sami, Kven, and Norwegian background. The next largest settlement, Haugnes, contained 10 primarily Sami households. That year, Friis notes that five Sami families herded between 500 and 600 reindeer to Arnøya.

Even though mid-19th century immigrants such as Johannesen and Mathisen may have temporarily increased and diversified the ethnic composition of Arnøy's population, the last decades of that century also witnessed increased emigration from North Norway, including moves to America.[28] This may explain the fact that by the 1900 census, Arnøy's total population declined (from 365, in 1875) to 236, then scattered between some 24 small settlements (RHD 1900). These fluctuating late 19th century populations appear to confirm the historic 'limits to growth' inherent in the *fiskarbonde* economy, natural conditions, and impartible inheritance.

By the 1900 census, the Norwegianization policy discussed above actually listed Finnish or Sami newcomers such as Johannes Mathisen as 'Norwegians,' even though most would have continued to speak their mother tongue and consider themselves Sami or Finnish in everyday life (Bjørlund, ibid). The 1900 census lists Årviksand as the island's largest settlement, with an astonishing 91 people. Nygard (now within Lauksletta) followed with 40, and Grunnfjord (now Arnøyhamn) with 36.

Early 20th Century Adaptations

Today and for generations, all land on Arnøya is privately owned and named. North Norway's decentralized settlement pattern meant that many farms stretched from the peaks of mountains (*fjellet*) down to the sea shore, to a point half way between the high and low water mark. However, in the few densely populated, nucleated settlements like Årviksand, farms were necessarily composed of discontinuous parcels, as is illustrated by one Årviksand farm, which operated until the 1970s. The farm *consisted of about twenty* ***mål*** *(or* ***dekar****). One* ***mål*** *equals one* ***dekar****, or 1000 m2. My parents owned twelve* ***mål*** *by the house, and another eight near the football place. Also, from the river, over 100 m wide, two or three kilometres long, to the top of the mountains.*

Traditionally, but to a lesser extent today, land is either purchased or inherited, normally by the oldest son from his parents. In turn, the son acquiring the land was obliged to care for his aged parents. Thus although people remember the *odel* system of inheritance, discussed above in relation to historic Arnøya, the older system is less relevant since the *fiskarbonde* economy was replaced by the

---

[28] See Aas (1998) for an account of late 19th century emigration from North Norway.

industrial fishery, as discussed below.[29] Even so, one informant explained the *odel* system as follows: *If there was more than one son, we have* ***odel****, or freehold property custom, in which, the oldest son has right to the property. If you have five sons, the oldest one, he will have the farm, and can use his* ***odel*** *rights to get it. No brother can do anything about it. He can take everything and do with it as he wants, in terms of making money out of it. But if it became an agreement that one of the other five wants to own some land, and wants to buy it from his father's farm, but he is not the one who has the* ***odel****, you have to make an agreement with the other four, and then he can buy it, from his father's land. Thus, the oldest son inherits; others must make agreements or purchase.* Siblings not inheriting land sometimes received money in lieu of land while those actually inheriting farm land often had to pay for it, as well as to look after their aging parents.

The same informant paraphrased two deeds (*skjøte*), describing land transfers occurring in 1907 and 1949. In the first, the informant's grandfather (Mikal) buys the farm from his father; in the second, Mikal sells the farm to the informant's father, Kåre. My informant pointed to the 1907 deed, translating it as he read, as follows*: My grandfather Mikal bought the farm from his father, Petter Kristiansen. And he [Petter] is selling his part of the paper [deed], which was registered in 1892, as farm 57, user number six. Mikal bought it in 1907 and it gets a new name: Sandaker. The price he had to pay is 50 NOK. It should follow the buyer's family for some years to come*. The 1949 deed records not only the land transfer but also specific *odel* conditions Kåre agreed to as a condition of purchase. *The people signed under, Mikal Mattiesen and his wife Kristianna, say that they have sold the farm to their son Kåre Sandaker. The farm belongs [in] Skjervøy County, it's called farm No. 57, user number 9. They sell it with house standing on the farm and all rights to the farm. My father [Kåre] had to pay 3800 NOK, back in 1949. The buyer should also, as long as his parents live, give them three living rooms, lights inside, and something to burn in the oven [that is, peat]. [The buyer should also] clean their house, give them food and clothes, and if they became sick in their older days, give them free nursing. And, finally, it says, this farm belongs to the buyer Kåre Sandaker, and his heirs. The price on what my father had to do for his parents, they put a five year price to bring them free house, and light, and so on to be about 2,200 NOK, or 400 NOK a year.* The man explained his father's obligations to his parents. *As part of my father [Kåre's] agreement to buy the farm, he also signed that as long as his parents were living, he agreed to take care of them. I think it is considered natural that the child who gets the farm should have responsibility for*

---

[29] By including descriptions of older realities, such as the *odel* or the *fiskarbonde* economy, I realize that some might accuse me of essentializing (or romanticizing) the peasant character of contemporary Arnøya. Instead, I suggest that it is precisely the shared memory and continued currency of such older, highly localized knowledge and customs, coupled with an increasing knowledge of the world beyond Arnøya, which constitutes the contemporary culture, and distinguishes it from the culture of urban Norway, a culture drawn from *diverse* rural regions (cf. e.g., Gullestad 1984).

his old parents. This allows the oldest son to go to siblings who get nothing and say: 'look I paid 3800 NOK for the farm and also agreed to take care of our parents.' Although Odelsloven and impartible inheritance attempted to prevent the division of the farmland so essential to the older fiskarbonde economy, pragmatic realities, especially since the recent emergence of the industrial fishery, reduced the relevance of such systems of land transfer and caused 'indivisible' farms to be subdivided (See also Saugestad 1988: 58). Indeed, subdivision eventually split Kåre's farm, described above. Kåre's brothers Petter, Bendix, and Maks all received house lots on what had once been Mikal Mattiesen's farm. Kåre's son said, *they cut Mikal's farm into three pieces. That one [pointing] is called Moberg (black heavy ground), Petter's share; Bendix got what he called* ***Thune****.*

In earlier times, other economic assets, such as *laksenot*, the customary places where salmon nets were set, peat bogs, and places for picking cloudberries (*moltebær)* were also inheritable and owned property. For example, before 1900, Johannes Mathisen of Lauksletta, whom we met above in connection with Finnish immigration to Arnøya, subdivided his land (between Klubben and Storelva) between two sons [Mattis and Johan], as a descendant explained. *It was divided between his two sons. What made it valuable were the eleven different places where salmon nets* ***(laksenot)*** *could be set. He obtained these through his marriage with Karin Larsen. Today, these berths are used by Gunnar and Frank. They have agreed that every second year, each would use these salmon places. Both can use them each year but they change where they fish (berths). Further, the one who can use the salmon berths can also pick the* ***moltebær****.*

All farms were named according to some natural feature characteristic of the property, and numbered.[30] Much as Gaffin (1996) describes for the Faeroese landscape, Arnøy locations, especially those outside of settlements have toponymic names (such as *hemmelig* [meaning secret place] or *storhammer* [meaning big boulder]), for practical purposes, as one man explained: *When people had sheep, everyone had sheep, and the sheep are on the whole mountain and then they get looking for the sheep, then you had small places with different names. They must have known every rock.* Toponyms expressed people's practical relationship with the landscape.

Customary law recognized boundaries between units of land and there also appear to have been localized boundaries between settlements, as one man explained in relationship to hunting and angling rights. *There was a lot of salmon in the lake near Årvik. There was a boundary across that lake. In the middle of Langfjord Pond there was a stone boundary between where salmon fishers from Årvik and those from Langfjord could fish. People from Akkarvik hunted* ***rype***

[30] Frykman and Löfgren (1987) trace the origins of contemporary middle-class Sweden back to various images of Swedish peasantry, when intimate knowledge of the natural landscape was necessary for making a living.

[ptarmigan] as far as Rundhaugen and Kistefjellet and Høgtindan. All communities went as far inland as the highest hills.

Just before, and especially after the Second World War, some of the peoples who formerly lived in small settlements around the island moved to communities such as Årviksand, transforming strangers into neighbours. Newcomers moving to larger settlements were either unaware of or unwilling to adhere to local custodial notions of responsibility over local resources such as a nearby river or berry patch, rendering such local knowledge and property rights less operational. My conversations with older people led me to conclude that the ecological consequences of crowding more people into less space over-utilized or depleted some local resources, because of ambiguities about just who should take care of local resources. Although this phenomenon of settlement concentration, with its attendant consequences for local resource management, likely occurred in each of the island's small villages, I heard most about the phenomenon in Akkarvik and Årviksand. In the latter place, for instance, people attributed the depletion of Atlantic salmon formerly endemic to the local river to the influx of families from other settlements following the Second World War. According to their explanation, (relative) newcomers failed to respect the older, local management regime. Similarly, Akkarvik's finely tuned custodial regime, which permitted local families living adjacent to the river to set nets across one-third of the stream, were ignored by newcomers, with damaging consequences for salmon.

Historic Socioeconomic Differentiation

Before the transition to today's industrial fishing economy, land was necessary in order for small-scale fisher-farmers to make a living. Some of those without land comprised a landless 'underclass' of cotters, working for landowners to whom they were not usually related.[31] In Ivar Bjørklund's (personal communication, 1995) words, historic economic differentiation meant that *the difference 100 years ago between rich and poor was enormous, but on the other hand there weren't many*

---

[31] I acknowledge Brox's (1965: 86) point that a household's children and its servants were, in many ways, "equal members of the household." However, Brox also claims that in his unnamed North Norwegian fjord "there were few or no exclusive territorial rights" (p. 84) and that capital accumulation in fishing was "almost non-existant" (p. 84). In short, fjordal society was highly egalitarian. Brox then writes that his informants perceived "marked differences in levels of prosperity between the households" (p. 85). The locals' explanation was that those 'lucky' enough to consume, rather than to have to sell, valued surplus agricultural products, such as butter, were prosperous. Nevertheless, Brox claims that the relative prosperity of some households was not rooted in capital accumulation. I find Brox's argument stimulating even though its logic appears circular. Thus, prosperous households were those able to attract and control a non-related servant labour force of fishers or farm workers. However, for Brox, households able to attract such labour did so by providing "generally appreciated goods" (p. 90), luxuries such as butter or veal, leading me to assume that such households had accumulated more capital and were more affluent than those forced to sell such goods. In short, contrary to his claim, Brox's data appears, like my own, to root prosperity in some form of capital accumulation.

rich ones! When helping me understand the 19th century censuses, Bjørklund explained two terms often appearing in these documents, *inderst* and *selveier*. The former includes those owning nothing, persons who were either young people just beginning their careers and lacking property, or old people who had lost everything. On the other hand, the term *selveier* refers to a person who owns property and is, in principle, economically better off than others.

There were several categories of *inderst* persons, including *pleiesønn* and *husmann*. *Pleiesønn* were children taken in custody and sometimes adopted by other, usually *selveie*r families. One woman described a *husmann* (or crofter) as follows: *If you are a farmer, you have a contract with some working men, who work on your land, but it's still your ground, for a term or life. The farmer owns the land; the* ***husmann*** *works it, and pays rent.*

One man told me that until around the Second World War, poor people would go visiting (*laupe*) between houses to obtain coffee and bread. *I heard it from my father. And this happened in Arnøyhamn. They were begging, just visiting. And the normal way of behaving when you get a visitor is that you get some coffee and cakes. At that time it was a small community and they were much more concerned about each another. And the wealthy people they knew that the other guy don't have much. So they don't make a big deal about it. They just cook some coffee and cut some bread, making the other guy not to know that they gave him bread because they knew that back home he didn't have any.* The same man then explained that although there are still poor people on Arnøya, nowadays they *get help from government. In the old days they called it* ***fattig kassa****, or poor box. And those people that had to go to the* ***fattig kassa*** *were much more ashamed of letting others know that you had to get some money from the social department than now. Nowadays, it's more or less normal; no one cares.*[32]

Making a Living

Until the Second World War and in some cases until the 1960s, most Arnøy families were fisher-farmers, a pluralistic economic adaptation described by the Norwegian word *fiskarbonde*. Men worked in seasonal fisheries and women in the home and farm.[33] This gendered division of labour and domestic mode of production differed in at least three respects from the present, industrialized fishery. First, families consumed more of what they produced than now, each act with its consequences. Second, the older economy was more seasonally ordered than that

---

[32] During the 1930s, but especially following the war in the 1950s and 1960s, a variety of new social benefits were extended to all Norwegians. Old age pensions began in 1936; after the war, health insurance was made obligatory in 1957, family allowances (*barnetrygd*) were added in 1958, disability allowances in 1961, and so on (See Danielsen 1995: 393).

[33] Studies on the gendered character of the *fiskarbonde* econonmy include Aubert et al, 1970; Saugestad 1988; Austrem and Ingebrigtsen 1992; and Elstad 1999.

which replaced it. And thirdly, the *fiskarbonde* economy required impartible inheritance to be viable. Apparently (and we can only assess this indirectly), agricultural holdings needed to be of sufficient size to work, meaning that only one child could remain on the family's land.

Informants remembered the *fiskarbonde* economy as follows: When I was my son's age (13) *we didn't have the road -- we only had the road from one end of the settlement to the other -- Haugnes to Elva. In the summer time we had to take care of the crops for the winter. My father was a fisherman and just came home to take care of the hay, put up things [wooden hay ricks, called* ***høyhesje****] to dry the hay on [called* ***såte****]. And then he did another type of fishing. My mother and the kids did the rest of the work. Nowadays they [i.e., children] don't know what work is.* Similarly, an elderly woman recalled the seasonality inherent in the historic economy. *We had two or three cows, and ten or twelve sheep. In the 1940s and 1950s everyone had sheep. We had four children. [There was] all kinds of work. Cutting* ***tare*** *(kelp) in November and December for food for the sheep. We would cut* ***tare*** *much of the winter. We could also cut and boil the heads of the cod for the sheep [cooked fish heads and kelp were called* ***løypning****]. We had a big oven [stove] in the sheep house. In spring, March, April and May, you have lambing. We had milk, butter, and cheese, for ourselves and sold some locally, especially to people who did not have cows. But a company also bought the milk and butter. Women's roles were very important in maintaining the farm.*

An elderly man recalled his mother's daily duties. *Her jobs included cooking fodder for animals, housework. We were seven children; I was second oldest. Mother arose at 5:00 AM; she seldom bedded until 11:00 PM, as after the kids were in bed she worked on clothing and other work. She also had to get the family's water, using a* ***tønne****, a wooden cask holding 120 litres of water. She hauled water on a sled about 1.5 meters long. The back section of the bottom of the sled had a ring seal bottom, with seal hair facing toward rear. When the one-skied sled went uphill the seal hair prevented it going backward. Households required one* ***tønne*** *of water daily while the barn animals required two. Mother did it every day, with help of two sons. The water source was about 100 metres distant. She would also agitate or rinse (****skylle****) her clothes in this water.*

The same man recalled the activities of his youth. *We would cut grass on the* ***lia****, mountainside, and bring grass to the bottom to dry it. Also we would cut turf in summer. Turf was burned in house and barn. In the barn turf was burned on a stove to cook grasses. The hay was too rough, and was therefore cooked in hot water, to make it easier for animals to eat. Fish heads were also cooked for animals to eat. When it was warm the animals liked it better. Cows, sheep, goats would all eat cooked fish heads. Cutting grass on the* ***li****, farmers would be able to predict the numbers of animals they could support for winter. More work, more animals, more cubic metres, more animals. In springtime, if snow was a few weeks longer than usual, people would cut* ***tare*** *(kelp). We would cut it from a boat. We also fed* *animals* ***brom****, new buds from small birch. We had a family of seven children, two cows, one ox, one calf, five goats (primarily for milking),(baby goats are called* ***kji****, were eaten after being fattened on spring grass), twelve sheep and one horse (used in the garden and in winter to bring cod heads from sea to barn for cooking for the animals. In winter when boats came in, and the cod heads were cut off, the horse was used to transport cod heads to barn).*

Women spun and knitted warm woollen clothes, including socks *(lest*), trousers *(ullbukse*), and others. They baked bread and prepared the common meals of fish, soups, and coffee on stoves (*ovn*) fuelled by peat, which also provided heat for the house. A predictable schedule of daily meals ordered daily life, a routine one man recalled as follows*: on Sunday, we had meat, from our animals or from birds from Fugløya [bird island]. Every springtime we would go to there to get birds and eggs. We'd salt the meat. We'd hang the food in a small house called a* ***stabbur****, a food storehouse. Dried meats were stored there. We would make a* ***kjøtt suppe*** *(a meat soup). That would be every Sunday. Monday, fish, salt fish. Sunday, you would take the nets up out of water, so no fish could be caught. Tuesday, fresh fish on Wednesday you have* ***steakefisk****, fried, usually* ***sei*** *(saithe). Thursday,* ***kveite*** *[halibut] or* ***uer*** *[red fish]* ***suppe*** *but you must have* ***tymie*** *(thyme) in fish soup. Friday, anything. Saturday -* ***Grøt*** *[or* ***ris graut****] rice porridge, every Saturday! In summer time, we have meat from birds on Wednesday. Also seal meat or* ***nise*** *(porpoise). Lutefisk*, cod soaked in lye, then rinsed in water, and cooked, is not as common on Arnøya as further south in the Lofoten Islands, although a traditional fish dish called *mølje* and sweet rolled pancake called *lefse* remain popular.

In sum, although each of these narratives discusses particular characteristics of the fiskarbonde economy, there are some overlaps. Among these are the repeated use of the word work, daily and seasonal orderings of time, the autonomy of men and women in their gender-specific roles, and a close relationship with the natural environment.

The Historic Fishery

Arnøy men sailed (later motored) each winter south to the important cod fishery around the Lofoten Islands. In spring, fishers sailed north to the Finnmark coast to fish for cod. While not the only species, cod was particularly important. *Sei* was second in importance, as this more localized fishery occurred primarily during late summer, after the harvest of hay was complete, and prior to the dominant cod fisheries.

Nowadays, some Arnøy fishers seine for herring (*sild)* in fjords south of Arnøya during fall, although most of the older people I interviewed had little to say about the historic herring fishery.

An older man recalled the annual cycle of fisheries during the early 20th century, his father's time. *Everybody was in the Lofotens in the wintertime. My father would go around the middle of January and come home about 20th of April, around Easter. He fished with Isak i Geitvika [Isak Mikkelsen, Geitvik]. In the old days they'd say Isak Geitvika. They [his father] would return (from Lofotens) here and then go to (Makkaur) Båtsfjord, Finnmark, fishing. They would stay in Berlevåg, some stayed in Gamvik. [Isak had a] 32 feet boat powered by a 20 horsepower motor. They'd go about five knots. He'd have eight men fishing with him. They would be back in mid-May; they only had one month in Finnmark. Then my father would start turfing. Then haying, from mid may till the 25th of July. Then he'd start fishing* ***sei****, in Sotnes; in Sotnesbåen there is much* ***sei****. Sotnesbåen is about 10 fathoms deep.* ***Sei*** *fishing began at the end of July. They'd fish for about one month, for* ***sei****, until about the 20th of August. They would then leave the inshore waters (like Sotnesbåen), then the fish go offshore into deeper waters and fishers would follow. Sotnes fishers would fish for* ***sei*** *at a place called Seibåen, more offshore. They would fish from the 20th of August until about the 20th of September. After the 20th of September they would fish from bigger boats (****fembøring****) until Christmas.*

Like the other men of his generation, my older informant began fishing at age 15, going aboard a fishing boat on the afternoon of his confirmation into the Lutheran Church. He then described some of the technologies of the Lofoten fishery. *I was fifteen years when I first went. They would fish for* ***torsk****, using* ***angel*** *(baited hook and line),* ***jukse*** *(jigger), and* ***garn*** *(gill-styled net). And today they have the trawl,* ***rundfisk trål****, and they used that. And they had some special place on the sea, in this place you can use trawl, in this place you can use* ***jukse*** *[this likely refers to 'co-management' the distinctive fisheries management system developed by Lofoten fishers.[34]] They start with angel and line (long line), and so they came to line and garn (net), and during the last 30-40 years, they have the* ***rundfisk*** *trawl. In 1940 they started using the* ***snurpe not****, (seine like net) for cod and* ***sild****. They started using it in 1940 but only used it for 10 years. By then they had taken up all the big fishes, and then they must stop, they caught too much big fish (****los****), the large fish which show the way (lead schools of cod).*

Memories differed on just how fish caught in the Lofotens were marketed, although most people said that it was sold fresh to Lofoten merchants. One man recalled*: When you go to the Lofotens, Isak i Geitvik would take all the crew in to Havnnes, to Giæver's (merchants in Havnnes, Lyngenfjord), and there they would take nets, food for their family back here, bring it here [Arnøya], then head to Lofotens. They would sell their fish in the Lofotens, take the money and head to Havnnes, to repay Giævers. At Lofotens they'd trade fish with Johansen Stamsund, he was the biggest, he built the school, church, houses, for fishers, at Stamsund, near Svolvær*. Similarly, fish from the Spring Finnmark fishery was sold locally in Finnmark, as one person put it, *at the nearest place*.

[34] Jentoft and Kristoffersen (1989), in their seminal paper, describe the historic development of co-management in the Lofoten fishery. Holm et al (2000) argue that co-management manages fishing space more than it manages fish stocks.

Elsewhere in northern Norway, probably including Arnøya, money earned in the Lofoten fishery was, by the early 20th century, important to household economies, particularly because many other seasonal activities, such as local fishing or turfing, produced goods consumed by households.[35]

In addition to the migratory Lofoten and Finnmark fisheries, people fished locally around Arnøya. An Årviksand man summarized the historic fishery from that community. *Spring and summer, they fished just outside here for* ***sei, torsk****, or* ***kveite****, using* ***jukse*** *(jigger), line, 400 hooks. In Fall, there was more work on land, turf to be cut, 6000 blocks [roughly 25-30 cm long], dried on the ground, left out over summer so water could drain out. Each family had their own turf places.* Although cod was the most important species overall, especially in the transient fisheries, the local (late summer) **sei** (saithe) fishery was important, as recalled by an older informant. *They would use a small, seven-fathom square net (i.e., seven fathoms, by seven fathoms), with one boat on each 'arm' corner. Fishing would occur on the tides. Every six hours there is no tide (tide or* ***tidevann****), the net would be placed on the bottom. When the tide began flowing, the* ***sei*** *would swim toward the tide and the net would be raised. They would use four small boats (****spisse****, a 4.5 metre row boat) with two people in each. Fish would be taken to the land, where it was placed on the* ***fiskehjell*** *(fish drying scaffolds). Once dried, they would bring it to the* ***handelsmann*** *(merchant). The biggest here was Brockmann, of Nikkeby (originally, most of the traders came from Germany and Holland).*

Some cod caught locally was salted for domestic consumption or dried on flat stones to be exported as *bacalao* (dried salted cod). But most of the *sei* and *torsk* fishers caught near Arnøya were split, hung wooden racks (*hjell*); air dried on wooden racks, and sold as unsalted stockfish or split salted fish (*kleppfisk*). Brooke (1823), who travelled to north Norway in the summer of 1820, described preserving *sei* and cod as stockfish. The fish were decapitated and eviscerated, then hung separated from one another on wooden scaffolds. Generally, in north Norway stockfish took three months to dry, and was best dried between February and June, as near to the Gulf Stream as possible, so as to reduce freezing. Brooke observed, *a fish once dried in the air in this manner will keep good for several years; and in order to insure this, a law of great importance to the fishery forbids, under severe penalties, the taking down the fish from the yells [****hjell****] before the 14th of June, when they are supposed to be thoroughly dried. Previous to this, the agent of the bishop of the Nordlands and Finnmark, who generally farms his tithes, goes around and ascertains the quality of each yell [hjell].*[36]

Arnøy people still produce some stockfish for domestic consumption, but had mixed memories about historic commercial stockfish production. One man said that stockfish production created a lot of jobs; another recalled that Arnøy stockfish was

---

[35] See Dyrvik (1993); also, see Hovland et al (1982) for a description of the Lofoten fishery.

[36] See Brooke (1823: 244).

Stockfish hanging on hjell, Rotsund 1995. (Photo by author)

of poor quality, perhaps as another man explained, *if the local merchants would not buy my [fresh] fish I would dry it as stockfish. When frost ruined the fish, we'd get minimum. This fish would go to West Africa.* Finally, and regarding marketing, a man recalled, *they [the Giævers, merchants from nearby Havnnes] used to come to Arnøya in July or August and buy the fish from small fishers.*

During the period described here larger boats replaced smaller ones. An Årviksand man described smaller boats once used in the local fishery. *Before the* ***molo*** *[stone breakwater] was here, people pulled their boats up at* ***stø*** *[landing places where were stones removed]. Four or five man crews, 20-foot boats called* ***kjeks*** *almost like a [double ended] Nordlands boat. They lacked motors in boats. Capstans were used. Once the* ***molo*** *was built bigger boats were used.* Another, older Årviksand man described boats used prior to construction of the *molo* that created an artificial harbour. *We had to put them on the land, every night. Twenty-five foot long boats, five horse power motors, inside, with a little house over it. Four or five men fished. We used* ***jukse, garn****, line.* One of these boats was the *fembøring* [from the noun *fem*, meaning five, and the verb *bo*, to live].[37] *Fembøring* often had a small house on it that could be taken ashore to sleep in. One old man recalled the *fembøring* used by his father, *The five brothers were Anders, Randolf, Ole, Olaf, Halvor.* ***Fembøring****, five people making a living from that boat. These boats measured up to twenty-six feet. It was a Nordlands-styled boat.*

[37] Another, less common, meaning of *fembøring* pertains to the vessel's construction, specifically that there are five wooden planks (*fem bord*) on each side of the boat. Some boats where once made locally, using, until the Russian Revolution in 1917, timber imported from Russia, as part of the Pomor trade. For a brief overview of the Pomor trade, see Niemi (1995).

By the 1930s, the necessity of fishing further from home led fishers to begin using larger boats. One retired fisher explained his sale of a smaller boat and purchase of a larger one as follows. *It [the Terningen] was too small. I had to go far on the ocean. I had to go to Finnmark and Lofoten. January, February and March and April around here. After April, to Finnmark.*

Merchants, Credit, and Cash

The narratives of Arnøy people were inconsistent about how fish and other local products were sold, and to whom, and through what medium of exchange. Yet they agreed on some points. All agreed that Arnøya never had a *nessekonge* [literally, the king of a cape or headland, who 'owned' the whole village], that is, a dominant merchant.[38] People usually smiled (or laughed) whenever I used the word *nessekonge*, for reasons I cannot explain. Perhaps it was simply hearing an English-speaking foreigner express an interest in this old Norwegian word, or that in the context of today's social democracy, my query about very unequal earlier times prompted uneasiness.

Although Skjervøy had had an historic *nessekonge* (see footnote 37), he apparently did not control Arnøya. As one man put it, *there was no* ***Nessekonge*** *on Arnøya. But Johannes Giæver [of Havnnes] was a* ***Nessekonge***. Another man described relations between merchants like Giæver and fishers, as follows: *the ones who produced the fish, that is, caught and dried it would have a credit relationship with* ***Nessekonge****, but the ones who delivered the fish [to the merchant] would get a little bit of money. You always owed them money. No matter how much you fished and paid them back, you always owed them. So more or less you were owned by these guys.*

The same fisher described his father's role in organizing Arnøy fishers in the face of merchants like Giæver. *[My father] Ingvald [born in 1897] was one of those who fought for the* ***Råfisklag*** *[union of fishers] by making the fishermen organized. Then the politicians agreed. So he got fishermen organized. He was among the first on Arnøya who fought for this. The* ***nessekonge*** *was Johannes Giæver of Havnnes. In Tromsø, one of these Giæver people was recently saying that they were very sorry for what they'd done in earlier days.* [John: Did your father deal with Johannes Giæver?] *Yes. There were no other* ***Nessekonge*** *near here. There was one in Skjervøy but he didn't control Arnøya. So Giæver controlled it.* Another man described fisher-merchant relations and compared historic production and consumption with the present. *You see the fishermen were gone for many months,*

[38] My interviews (See Appendix 1) failed to unravel the role played by *nessekonge* in the historic economy. Twentieth century merchants Brockman (of Nikkeby) and especially Giæver (of Havnnes), whose firm continues, were mentioned from time to time, but seldom in ways suggesting a distinct system of merchant-fisher relations, in which merchants supplied fishers on credit, such as in Newfoundland. Neither the 16th century Skjervøy merchant Christen Heggelund (Johnsen 1996:11) nor the pre-World War One Akkarvik merchant mentioned by Smith (1997) were mentioned.

Havnnes. Lyngenfjord, 1995 (Photo by Karen Olsson)

to the Lofotens, and those who were home, they got credit, recorded in the book. So when the fishers came home, they paid, maybe two times a year. Everyone bought on credit. In winter, all men were gone to the Lofotens. The possibility for the shops to have credit was more flexible than today. People produced more of what they consumed and didn't waste money at the shop. An Årviksand man added the moralistic connotation that credit was restricted to poorer fishers, proudly recalling that he only acquired *not* (fish seines) and *garn* (fish nets) on credit. In his words: *I usually sold fish fresh to Amandus Pettersen [of Årviksand]. Sometimes I sold it salted. When we had a lot of fish, perhaps seven or eight thousand kilos, then we'd go to Tromsø. But ordinarily I would sell to Amandus Pettersen. We had money. We paid for our food. I never had credit in my time but some others here did have credit. I worked hard. But when I bought* ***not*** *and* ***garn****, I took these on credit, paying at the end of the season. But not on food.*[39]

Another man told me that in the old days *when people couldn't pay, they [merchants] kept [took] their ground [land]. That happened on Nord Fugløya (North Bird Island).* Although I suspect that such expropriation was unusual, it appears that merchants could take a fisher's land or other economic assets as payment on an outstanding debt, as also happened in historic Newfoundland.

---

[39] An older informant suggested that merchants supplied fishers on credit until the 1920s and that after that; cash was more commonly used.

Young man's confirmation. Årviksand 1995. (Photo by Karen Olsson)

Religion

Arnøy people belong to the Lutheran church and certain religious rituals, such as confirmation, remain important. However, my sense is that religion is less important today than in earlier generations, when *Læstadianism* shaped social action and morality, insulating adherents from the corruption associated with the outside world. *Læstadianism*, a fundamentalist sect within the state Lutheran church, began in the mid 19th century as a millenarian movement initiated by Swedish priest Lars Levi Læstadius (1800-1861).[39] The Sami and Finnish peoples who embraced Læstadius' fervent, emotional, and conservative teachings did so because these condoned the impoverished living conditions both peoples then endured. Moreover, in the context of the modernization policies then proffered by the Norwegian government, policies aimed at assimilating the 'backward' Sami and immigrant Kvens (Finnish) into mainstream Norwegian society, *Læstadianism* offered the symbolic tools with which to construct what Bjørklund (1988 --- following Schwimmer) calls an 'opposition ideology,' inverting the dominant metaphors of Norwegian modernization policy into local realities, and defending the moral virtues of Sami and *Kven* language, family, and community. Through a complex combination of reindeer herding routes and bilateral kinship relations, *Læstadianism* diffused from its interior base to three coastal districts where Sami lived, and one of these was Skjervøy.

[39] References to Læstadianism in this paragraph are listed above in footnote 20.

The Lutheran priest at Skjervøy, whose grandfather was Sami and fluent in both Samisk and Finnish, described *Læstadianism* as follows: *Læstadianism is a part of Lutheranism. Læstadius was a Lutheran priest. The people, who were awakened by him and took his gospel to Norway, were Lapps. They were filled by Læstadius, by his very strong and clear message, and very demanding preaching. 'That's wrong, that's right.' He was charismatic. This movement was a very spiritual movement. They know what to believe. They said, 'we are a part of the Lutheran church in our land'. This movement was one big movement and Læstadius was the prophet.*

On Arnøya, Læstadianism is less important now than a generation ago. As recently as the 1960s, however, Brox (1972) reported a strict Læstadian moral code at 'Sand,' one of nine North Norwegian communities he visited. Today, people recall the era Brox describes with recollections of how Læstadianism once ordered local social life and morality.A young Arnøy man, whose grandfather was a Læstadian lay preacher on Arnøya, explained why the sect became popular in the late 19th century.[40] *They [people] needed moral speaking; there was much drinking, and so on. There was much poverty, small houses, plenty children, small boats, husbands away fishing, wife and kids taking care of the farm. Everything was small [scale], you have, maybe two or three sheep, a goat, some fishing nets. And to forget, they drank. Again you have so-called* ***Nessekonge****. They have a shop; they get cheap booze from Denmark. Norway was a part of Denmark then. They [fishers] brought booze, came home, got drunk, beat their children, it was a hard time, I believe.*

Next, he described his own Læstadian upbringing, his grandfather's role in encouraging participation, the fundamentalist morality prescribed by the sect, and his eventual withdrawal from meetings. *I remember growing up; he [grandfather] didn't allow us to have a Christmas tree. I remember visiting other homes and they'd have a Christmas tree and we just have the gifts lying in a chair. I remember I felt sorry. I was born in 1952 and he [grandfather] died in 1958, in the autumn. Christmas trees were a sin. In the bible they told about the people that was walking in the desert and they made a calf out of a goat, and they was dancing around the golden calf and prayed to it. And they compared the Christmas tree to the golden calf. Læstadianists believed that by dancing around the Christmas tree you were praying to it. In the extreme part of that religion, they don't allow people to have flowers in their windows. They don't allow curtains, as it showed you have money to buy them. By having flowers and curtains, you were displaying wealth. In modern times, the radio was compared to the devil. When the television came, believers would not enter the house. It was not God's will to have a television. Too much from the outside world restricts one's ability to serve God. Drinking is not allowed, and no sex before marriage. If you lived with your spouse before marriage and he/she died, they are strict about remarriage. Divorce is not allowed. There are*

---

[40] I include this young informant's recollections as they describe a morality common to Læstadian areas long before today.

still a few Læstadians here, a few. I can remember when I grew up, in the springtime about Easter, and at Christmas time, priests from outside held big meetings attended by the whole Island. Like revival meetings, lasting two or three days. These are called Samling [assembly], meaning coming together. They occurred here in Arnøyhamn, in the schoolhouse, we didn't have a church house at that time. On the last day, they would begin about 2:00 PM, preaching, singing, etc. then stop for coffee and food, and recommence. In the evening around 5 or 6 came something called ***rørelse*** *[catharsis]. The listeners were moved to tears, they were hanging around each other's necks, and praying for forgiveness if they had done something wrong to someone else for the year. Oh the crying! I remember when I was crying, and begging for forgiveness. There was one old lady, a widow. She was begging my father for forgiveness because she had, in the springtime when there was plenty of snow, there was no grass for the sheep, so she had to go down when there was low water, and picked up some* ***tare*** *(kelp) and* ***tang*** *(seaweed) and did so without asking for permission. And I could not understand why that particular thing was something requiring that one cry and beg for forgiveness. So, my feeling at the time was that the whole thing was like a circus, a mass chain reaction. For me, as a child, I remember watching with open eyes. I remember when I came home, I told my father, that's it; I'm finished, no more for me. I did not want to go to meetings any more.* [John: what did your father say?] *I think he also came to that point, you know his father [a Læstadian preacher] was dead. And he don't have to show his father that he's going to the meetings. And maybe he had the same thoughts I had that day. During his last years of living, I cannot remember if he went to more of those meetings. That was one of his last. But they used to have at that time, during Sunday afternoon, they come together, ten to fifteen people in one house, and they have a sort of service, a simple speech, and they sing some hymns and drink some coffee, sort of a social meeting together and read the word of God. My mother was involved, to the extent she was married to my father and he was Mikal Mattiasen's (the preacher) son. So they expect him to go. She no longer goes, very seldom. Now we have [the official, Lutheran] church and people go there. The old Læstadianism is dying out. Now, there is a long time between meetings. I don't think they have one meeting each year, maybe one every second year.*

The man further explained his childhood attraction to the meetings and his eventual withdrawal from the sect. *As children we knew that when they have those big meetings, there would be plenty of cocoa, and plenty of food, good food. We looked forward to them. It was not every home that could afford to have cocoa. And most of those meetings were in wintertime and it was cold; you could go inside and have cocoa. I finally stopped going when I was eleven or twelve [i.e., 1963 or 1964]. By that time, we were told we had to be quiet, sit down and listen. It became a burden. You have to follow us to that meeting. When I reached that point, I say, 'what am I listening to, what's happening?'*

Lutheran Church and graveyard, Arnøyhamn, 1995. (Photo by Karen Olsson)

An elderly Akkarvik man underscored the importance of Laestadianism to local social life even though he claimed that his family did not participate. He said, Everybody *here was Læstadianer.*[41] *Meetings occurred in houses, every Sunday. They met.* [John: Did you participate in this?] *No. No.* [John: Did your parents participate?] *No, my father was from Fugløya, and there was not Læstadianism there. Fugløy people were more free. Læstadians don't like curtains, 'you shall never show yourself, you should stay back.' No radio, no television, these are not necessary for living. These things could be bad for people.*

The 1960s, a watershed decade for so much of Arnøy life, challenged the moralistic foundations of Læstadianism. Individuals and families began to withdraw from Læstadianism, as remembered by one Årviksand man who was a child at the time. *When I was a kid there was no Christmas tree in the school. A very bad thing that. Television also was thought to be a very bad thing. When we were little kids we had no Christmas tree, only when I was six or seven did we have one.*

A final example, from a Lauksletta man, illustrates again the stern moral principles of the sect but adds that Læstadian meetings, albeit with ever fewer participants, continue. *Alcohol was considered a 'sin'. Earlier, you couldn't drink in a house. Then you went outside, just around the corner and drank. (Same for dancing) you went outside, just around the corner, than you danced....Some of the people today don't have Christmas trees, don't have television, it's a sin.*

[41] In fact, most, though not everyone in Akkarvik, were Laestadianer.

The voices quoted above show that Læstadianism once set many of the social and moral standards of the community, uniting adherents in the celebration of family, sobriety, and community. Even though some believers, mainly older people, occasionally gather for Sunday afternoon Læstadian services at a house in Årviksand, their numbers, like the importance of the sect, are continually declining.

In describing the natural environment, the *fiskarbonde* economy, historic fishery, and older moral order, the voices aired above set the stage for changes increasingly common around the time of the Second World War, to which we now turn.

CHAPTER 2

THE BEGINNINGS OF CHANGE

Entitling this chapter as I have might lead to the erroneous impression that little change occurred before the interwar years and, obviously, this is not the case. Yet the years prior to the Second World War and especially the war itself were times of challenge and change for all Norwegians, including the people of Arnøya. In this chapter, people recall these years, and along with the accounts in the previous chapter, allow us to better understand the background for the more contemporary narratives presented later on.

The Interwar Years

As in much of the world, the interwar period began with the arrival of an unwelcomed visitor: the worldwide pandemic of 'Spanish' influenza, which arrived in 1918. The oldest man I interviewed remembered, *one of my sisters died in the 1918 Spanish flu sickness. Seven or eight younger people in Årviksand [died]. I don't know where the flu came from. Three of my nieces and nephews died, they were in their twenties. There were cases in other places, like Arnøyhamn. It occurred throughout Norway.*

Coincidentally, the price of fish in North Norway's fishery dependent economy plummeted during the global economic depression, primarily because Norwegian fish was largely exported to poor countries whose own agricultural exports also suffered during the depression, and because of an over production of fish by other North Atlantic producers.[1] One man recalled some of the effects of these difficult years. *In 1920s and 1930s it was difficult in Årviksand, difficult after the First World War. There was rationing during the First World War, rationing of bread, butter, milk, controlled [distributed] by Skjervøy kommune.* Another man put it this way: *The twenties was a bad time, all over Norway. Bad prices for fish. Five øres per kilo for fresh fish. In 1930, I was in Lofotens fishing, and it was still poor. In 1932, I started to work with the Staten [the state] building the molo [in Årviksand]. I was paid 65 øres per hour, when I started. I worked there until I retired, when I was 68 years old.*

Despite low fish prices, construction of a stone breakwater (*molo*) in Årviksand created an economic boom in that community, softening the negative consequences of the Depression. Some locals worked on the *molo* but the scale of the project also required transient, specialized labourers.

---

[1] A good account of the Norwegian fish trade during the Great Depression is presented in Apostle et al, (1998, 33ff).

Construction of the *molo* had other, lasting consequences, influencing much of the community's pre and post World War Two fortunes. The same man just quoted, claimed that actual construction began *in 1933. I was the first man and laid the first stone. The people who built the molo worked between 1933-69. Those from elsewhere boarded. There were 30-40 men. They worked from April to October. They had a barracks. They had two cooks. We used our hands. We had to dig five metres in the mountain to get stone.* Between the 1930s and the 1960s, a narrow gauge railway carried huge boulders from the stone quarries across the Årviksand river north to the work site. Notwithstanding the difficulties of the depression, this period also witnessed the physical expansion of Årviksand village, an expansion that the informant just quoted did not attribute to construction of the *molo*, but which I suspect was a consequence. Early in the 20th century, there were no houses on the north side of the river now bisecting the community. As my informant recalled, the community increased spatially and demographically in the mid-1930s. *In Årviksand, until 1920, no one lived on the north side [of the river]. Three persons (Karl Kristiansen, Christian Johansen, and Jørgen Pettersen) owned the north side. They bought the land on the north side of the river from the Kommune in 1905. These men had many sons. In 1934-35 they started building houses on the north side. This continued until 1944.* One couple that I interviewed were among those who settled on the north side of the river before World War Two. The husband recalled his childhood home in the now abandoned settlements of the east coast of Arnøya.

Anton Antonsen. He was my father's father, born in Tobbevika. My father's name was Harald. They used both [surnames] Antonsen and Ottesen. My father moved from Tobbevika to Sør Rekvika, because there was more land in Sør Rekvika. I was born in Sør Rekvika. We were married in 1938. We built a house here [in Årviksand] in 1939. His wife explained how the new couple acquired land on the north side of the river. *[We got the land] from my father. We had to work on the land as it was in the forest. My father was Einar Magnus Jørgensen. He lived near the river. This was his outfield. We own 50 mål.*

During the interwar years, Årviksand continued to be served by a general store named after its owner, AM Pettersen. Amandus Pettersen and his wife Jenny took in a foster child named Aksel Larsen [who we meet again in Chapter 3], who later took over the firm, and whose descendants still own (but no longer operate) the store.

At nearby Grunnfjord, another man began a shop during the interwar years. His son recalled: *In 1925, my father (Nils Borch Michalsen) started with a shop, post office and expedition of boat [boat business]. He was 25 years old. He came from Arnøyhamn. Nils Borch went to trades school in Tromsø.* One person claimed that in 1931 Grunnfjord received its post office and that its name changed to Arnøyhamn that year.

Similarly, at Lauksletta, the Nygaard family would, through a combination of inheritance and initiative, become prominent residents, eventually operating, in the words of a young family member, *a farm, shop, fishing factory, and every summer we'd fish wild salmon. Always we had something to do.* The family's rise to prominence began in the 1930s, when Johan Nygaard, who had worked for shopkeeper Fredrik Aas [eldest son of Sivert Aas, storekeeper], took over Fredrik Aas's firm in 1939. According to one of Nygaard's children, *[Fredrik's] wife asked our father to help and then she offered him to take over. It was not the only store in Lauksletta at that time. Skarbruket had a store and there was a cooperative store, run by an old teacher. It was opened in the evenings. There were many more people then [the 1930s], about 300. And you had people all around the other side [Laukøya]. There were seven or eight boats over 60 feet, and all had crews from Lauksletta or Nikkeby. After the war most came back, and had children,* ***and it's their children who moved away*** *[my emphasis]. Those who came back after the war are now about 80 years old.*

In 1912, a school was built at Haugnes, an exposed cape between Akkarvik and Arnøyhamn (Bentsen 1985). This middle school (*realskole*) accepted pupils from Akkarvik, Langfjord and Arnøyhamn. At the time, Lauksletta also had a school, and there were also two schools on the neighbouring island of Laukøya, one in Nikkeby, built around 1905, and one in Hellnes. The school at Årviksand attracted students from smaller, outlying settlements, along Arnøy's east coast, and northwest to Nord Rekvika, as remembered by an Årviksand man: *we had school in Martin Pettersen's [b. 1853] house. The children from Sør Rekvika to Nord Rekvika stayed here, going to school. They boarded with families. Seven or eight weeks in the spring, and the whole winter. The school operated the whole winter, but was divided into three-week sessions, with different children attending each session.*

Several people told me that from the 1930s-1950s, Akkarvik was the largest settlement on Arnøya. One man, whose family moved to Akkarvik in 1926, counted families and estimated that *around 1925 to 1926, there were around 20 families (with between 7-10 per household) for a total population of between 180-200, between the bottom of Akkarvik over to Sotnes.* He continued by explaining that when his family: *Applied for land in Akkarvik, there was no land available near the water, so he [father] had to move inland, and he bought 300 mål of land. All land around the water was then taken up.*

One Akkarvik man recalled that the population mushroomed in 1937 when people from various places established new farms. Akkarvik's excellent natural harbour provided safe anchorages for the larger vessels that were beginning to be used at the time.

The Second World War

Perhaps more than any other event of the 20th century, the Second World War was, as one man put it, a *boundary in time*. For older people, memories of the war are indelible, and for all, stories of Arnøya's heroic resistance continue to infuse pride in place.

There are several reasons for the war's importance. Norway was occupied by German troops from 1940 to 1945; the Norwegian government escaped to England; the Norwegian resistance, aided by Russian agents on Arnøya, monitored German convoys and troop movements; the population of Arnøya was forcibly deported from the island during the winter of 1944-45; and all buildings on Arnøya (except one) and in all communities north to the Finnmark-Russian border were burned by retreating German troops in the autumn of 1944.

The war continues to stir emotions, and to link remote Arnøya with the wider world. As I noted during 1995, a mixture of veneration and excitement characterized the 50th anniversary ceremonies both at local ceremonies at Årviksand, and at those I attended at Kilpisjävi Finland (the triple border where Norway, Sweden and Finland conjoin, and where, on 27 April 1945, German troops withdrew from Finland, and ultimately southward).

Germany invaded neutral Norway in April 1940,[2] battled with Norwegian and allied units until June, and then occupied Norway until 1945, fostering a program of Nazification which met with increasing resistance from Norwegians. Although the Germans had no base on Arnøya, they maintained stations at nearby Fulgøy Fyrlykt (North Fulgøy lighthouse), at the Engneset lighthouse near Skjervøy, and at Spildra, in Kvænangen Fjord.

An Arnøyhamn woman recalled resistance early in the war: *most [local men] were fishing in the Lofotens [in 1940 and 1941], they chose to go into the war. They went to English war ships, from their fishing boats. My father had to stop fishing because three of his crew went over to the English boats. Many never returned.* [John: I thought people from here took part in the resistance?] *Those were ordered to do so. [In 1940] six people from Arnøyhamn got a radio message that they had to come into Lyngen[fjord] and Kåfjord to be part of the fighting. That was after they had come home from fishing. When the boats arrived back from the Lofotens after 3 weeks fishing, the parents of those in the boats, those who didn't see their own children, started crying. They knew that something had happened, that they had gone to join the English war effort. Most were about 20 or 22 [years old].* I then asked her if she ever saw German troops. *I saw German soldiers but none were stationed here. There were German soldiers that came into the houses and investigated everything. Maybe to find radios, or something. And in the evenings you had to have black curtains so no light [then lantern light] went out. I was only eleven or twelve. We were afraid. And the old people were sceptical [cautious]. When we heard planes we had to hide. There was no close relationship.*

---

[2] As in World War One, Norway declared its neutrality when the Second World War broke out, in September 1939 (cf. Popperwell 1972).

We heard about the thing in Årviksand [see below], so we were very afraid. One man painted a somewhat different picture of life on Arnøya during the occupation. Conditions were, as he put it, *not so bad, we had food enough, potatoes, fish, animals. The Germans did not interrupt fishing. They told us to go to the Lofotens, to Finnmark. Get fishing! The Germans wanted the fish.*[3] *In 1942, I was in Reine, aged 18, in the Lofotens, fishing. I was in Gamvik, on Fugløya, when the war started. I was in a fishing boat with my grandpa, father of my father.*

The same man described how gossip enabled the dissemination of local and international news, notwithstanding the German ban on radios. *They took all radios, but we knew everything. I knew everything about Årviksand, from mouth to mouth. My uncle Julius, he was from Tobbevika (just north of Sør Rekvika). So Tobbevika people traded at Årviksand and had school in Årviksand. My friend Osmund was in school in Årvik, I was in school in Haugnes (until age 16). So my friend and I exchanged news.* [John: But wouldn't you be exchanging old information?] *Yes, but we were used to that, one month was okay, one week was one day. German propaganda told about their war effort, in Russia, convoys from America, bombing London, and so on. From the English, you heard their propaganda. So you heard from the Germans and the English, and you take half of it!* Another man described exchange with the German troops, as follows: *I visited the Germans and got some tobacco. The Germans didn't live near us. I took with me some dried meat I had made and the Germans gave me tobacco.*

Notwithstanding occasional instances of exchange and good will, the realities of occupation and war were unmistakable. Two unconnected examples of local resistance occurred during the war, one near Årviksand and the other at Nikkeby on the nearby island of Laukøy. Both, but particularly that near Årviksand, still endure in peoples' memories.

In the first case, in March 1943 two Norwegians were landed by boat, and one Russian radio operator, from submarine, at Moldfarvika to set up a spy station in a cave near Rotvåg, about six kilometres northeast of Årviksand (See Figure 2). Over the next number of months, these spies, supported by sympathetic locals, monitored German shipping. The location of the Rotåg spy station was concealed by radio signals coming from the Nikkeby station, described below. However, the Germans eventually located the Rotvåg station and on 25 August 1943, 200 German soldiers landed to investigate. The Germans forced a local man, Alf Jørgensen (variously remembered as the Devil, a paid German informant, or as simply unreliable) to lead them to the secret cave.[4] En route to the cave, Jørgensen bolted, fearing German or

[3] This is one of few references I heard to Germany's long- term plan that Norway would become the supplier of fish protein to Europe, the plan described in an excellent paper by Finstad (1999).

[4] The two Norwegian spies had established relations with Alf Jørgensen (see below) the previous year. The Rotvåg incident is the subject of Fjørtoft's (1981) book.

local retaliation, hid amidst boulders near the sea, and escaped that night by boat to Sweden.

Several days later, German troops conducted a house-to-house search at Årviksand and rounded up suspected sympathizers. Troops then forced a Rotvåg woman to lead them to the cave. Nearing it, she managed to jump in, and along with two male accomplices, shot some twenty Germans before being killed. The Germans detained suspected local sympathizers, whom they later took to Tromsø. Eventually, sixteen Norwegians (mostly from Arnøya) received prison sentences (either in Norway or Germany); another eight (seven from Arnøya) would be executed in Tromsø. Alf Jørgensen eventually returned to Årviksand in ignominy, to fish and to drink. Given the enormous human costs of this incident on Arnøya's small population, there are other bitter scars that continue to stigmatize some persons and families; these scars belong to the community.

The second incident involves Nikkeby native Hans Magnus Hansen, who, following a career in the Norwegian military, established a clandestine British wireless operation codenamed 'MU' during the war at Nikkeby.[5] The Germans knew that signals were coming from Laukøya but, remarkably, never found Hansen's station. Hansen's intelligence helped British RAF squadron 617 to sink the German battleship 'Tirpitz' on 12 November 1944, near Tromsø.

In November 1944, German forces began retreating south, burning virtually all buildings from Lyngenfjord north to the Norwegian-Russian border, hoping to deny haven to advancing Russian or other allied forces. The only house not burned on Arnøya was at Singla; the Giæver's historic Havnnes (Lyngenfjord) business also escaped Germany's scorched-earth policy. Just prior to torching North Norway, the Germans deported the local population to communities further south.

An older man recalled the deportation and the burning of Arnøy villages. *In November 1944 Årvik was burned. [The] Germans were moving south; some [deportees] went to Karlsøy. Three or four families went to Karlsøy. They moved everything, children, old people, one cow; they took everything they had. They moved just a few days before they burned all, in November 1944. Others were moved to the south of Norway, to Burøysund, Vannøya, the same day we were moved. We waited to be moved, and kept waiting, but we thought that Germany would go back [retreat]. The boat's name was Turid Oline, and they [laughing] changed the name and called it Noah's Ark, as we had a cow, people, and everything. We were depressed, and uncertain how long we'd be away. We were a little angry.... We were only one winter away, and came back in June and July. We lived in a big house in Vittnes, 3 or 4 families in that house. And that house is now in Tromsø now, it's a museum today, it's called Kvitnes gård.*

---

[5] I thank Carl Johan Bentsen for paraphrasing Hansen's 1992 diary, entitled *Fra Overvintring til Stasjon 'MU' Hemmelig etterretningstjeneste og opplerelsen under Krigen*, edited by Rolf Theil Endresen.

A Lauksletta man recalled, *during the war, my mother's family came from Nikkeby, and were moved to outside of Senja; they had kin there. My father's family were moved to Harstad, Kvæfjord. They had family there.*

As the reference above to 'Noah's Ark' shows, people somehow found humour in what must have been emotional devastation. One Årviksand woman laughed as she recalled leaving the community by boat as houses were ignited. *We could see when the house was on fire. Only the outhouse was not on fire. It was the only thing left!*

Once exiled from their homes, deported Arnøy people were on their own. One recalled, *[In Kvitneshuset] we had rationing, but we had to make some of our own food, by fishing.*

On 7 May 1945, Germany surrendered and a month later, Norwegian King Håkon 7 returned to Norway after a five-year exile in London.

Etter Krigen

With the end of the war in the spring of 1945, Norway and the rest of the world entered the post war period, or *etter krigen*. Arnøy people deported from their villages the previous fall returned to the ruins of their burned homes, outbuildings, and wharves. The next several years would be a time of rebuilding. One man remembered, *We had to find something to build houses with. I contacted Karlsøy kommune and asked to take the German house. I got that one and built it up the first winter. The war was over. We were happy to return, but we had to rebuild.* Some buildings were initially rebuilt temporarily from surplus war materials and there would later be a state-supported rebuilding program. One Årviksand family even received a house from the Government of Sweden. In 1995, its older occupant explained, *it's from Sweden, given as a gift. The house was prefabricated, built in Sweden. [My husband] built the second story.*

Central to our story is that not all settlements occupied before the war would be reoccupied after the war. Indeed, the period immediately *etter krigen* was not only one of rebuilding but of repopulating the landscape. I asked a man [the perhaps obvious] question of *why* some families did not return to their former settlements, some of which were small. No one resettled Tobbevika or Pankekeila because, as one person put it, *They were the smallest of places. I don't know why some chose not to rebuild these places. Sør Rekvika and Tobbevika people were old and went to Skjervøy.*

During the next two decades, new schools, fish plants, wharves, and roads would be a very visible part of the reconstruction of Arnøya, and an indication of just where people would now live and work. In 1946, for example, a makeshift boarding school (nicknamed 'the barracks' or *skole brakke*') was assembled near the bottom of Grunnfjord or Arnøyhamn, replacing the more substantial interwar years boarding school at Haugnes. The 'barracks,' was a simple wooden building, largely composed of discarded German war materials, and raised off the ground on wooden posts. The barracks operated between 1946-56, accommodating students from Langfjord to the eastern side of the Arnøyhamn bridge. The Arnøyhamn bridge was constructed before the Second World War. Once during the early 1950s, when ice destroyed it, day students from the 'other side' of Arnøyhamn had to board at the barracks. A more substantial bridge, part of the island's new road system, was built in 1968.

Over at Akkarvik, a temporary school (*provisorisk skole*) was built in 1947, for grades 1 to 7. A grammar school was built in Årviksand in 1950 and expanded around 1969. Until 1966, students from Årviksand completed grade 8 in Årviksand, and after that time, had the option of taking grade nine in Arnøyhamn. One student who completed grade 8 at Årviksand in 1966, recalled that enrolment numbered around 100 students at the time, some students came from as far away as Nord Rekvika, and that the school operated on a schedule of three weeks in school, and three weeks off. As of 1969, the Årviksand School had a student population of around 70 students. In 1958-59, a new school was built at Lauksletta that was expanded to include an internat, or dormitory, opened in 1970.

During the late 1950s and 1960s, government provided financial incentives (the amount I heard was 18,000 NOK per household) to households in smaller settlements to relocate to larger centres. Consequently, the last people to occupy Geitvik moved to Akkarvik in 1964; Sør Rekvik was abandoned in the early 1970s; Litlevika in the 1980s, and so on.

The Rise and Fall of Akkarvik

Placement of schools and other facilities had consequences that still arouse emotions, especially for those who believe that central policy makers overlooked their communities. Akkarvik is such a community. An Akkarvik (Akkarfjord) man told me that, *during and after the war the government in London had big plans for Akkarvik. It was supposed to be a centre for this region. I have seen the drawings and the plans.* According to some, the Norwegian government in exile secretly planned to develop eight big centres for North Norway, three of which would be in North Troms County. I could not, however, find confirmation of such plans. The Akkarvik man just quoted emphasized that Akkarvik had had a sizeable population prior to the war, as was noted above. He said: *there were a lot of people living here before the war and people did not just move here after the war.*

If Akkarvik were slated to become a centre after the war, such plans would have made good sense. Akkarfjord is the best natural harbour of any Arnøy settlement; the fjord faces south toward the mainland, and is generally protected from strong winds [some winds coming out of Lygenfjord can affect Akkarvik] and severe storms. And even though authorities would eventually favour the build-up of Årviksand and Arnøyhamn, Akkarvik boomed immediately following the war.

Two proximate developments explain this boom and another, namely Akkarvik's school, was more a reflection of the community's growth than a cause of it.

First, between around 1946 and 1958, Akkarvik became the retail 'capital' of Arnøya. Shops included a building supply store which sold materials used in the reconstruction of communities burned by the Germans in 1944. Goods were shipped to Akkarvik from as far south as Trondheim, then barged to rebuilding sites throughout the wider region. One man described the building supply store as: *a big shop where you could buy everything. North Troms and the Western part of Finnmark were built up [after the war] with materials from this shop. Høydal was the man who started the shop, in 1946. There were 6 people employed there, and then there were people working at loading and so on.* In addition to the building supply store, there were also food stores and a bakery, described by one woman as follows: *[The Akkarvik shop was] a lot bigger than the shop at Årviksand. The same man, Høydal owned the large shop, and there was also a bakery, owned by Stangland. There were two food stores; Aksel Johannessen owned one. The building supply place had food, as well. Aksel Johannassen (and after him, Asle Mikkelsen) ran a [general] store [which] closed around 1993.*

Second, Akkarvik's post war boom was supported by a succession of fish plants, described in the next chapter.

Finally, in order to educate the youth whose parents were attracted to Akkarvik by one or more of the developments just mentioned, in the mid-1950s a new school replaced the earlier temporary school built in 1947. By 1955, one person estimated that the new school [and village] had *almost 60 children at school and 55 houses with postal addresses. Today, there are only two children from here at the school in Arnøyhamn School. In 1955 the total population was about 200-300.* (Another person estimated that between 450-480 people lived in Akkarvik in 1955, with 82 youngsters enrolled at school). A third estimate, closer to that quoted above (and compatible with the official number of 56 for the 1958-59 school year)[6] was that, *after WW II and up to the mid 1950s, there were about 56 pupils in Akkarvik, it was the biggest school on all Arnøya. [But then] a lot moved out, to Skjervøy, to Tromsø. When there were 56 students, there were between 150-200 people in Akkarvik.* An Akkarvik couple remembered that the school opened in 1955: *The school here was* ***not*** *finished until 1955 and was conducted in various places. There was a temporary school here before and after the war. Our oldest daughter started grade one in the new school in 1957.* One couple whose family moved to Akkarvik in 1955 from Karlsøy explained why the new school was needed. *It (Akkarvik) was a big place at that time. Lots of people, and you could buy this house, with a boathouse, and everything was there.*

[6] Kommune school statistics for the 1958-59 year actually record 56 pupils at Arnøyhamn, Årviksand, and Akkarvik. (Skjervøy Kommune 1960).

Census data for the period from 1960-2001, presented in Table 2, for Arnøy's four villages and for the kommune centre of Skjervøy suggest that Akkarvik began to lose its population around 1960, just as Årviksand and Arnøyhamn began to increase.

Table 2 *Populations of Arnøya and Skjervøy, 1960-2001*

Year	Akkarvik	Arnøyhamn	Lauksletta	Årviksand	Skjervøy
1960	397[7]	263	223	188	1227
1970	123	244	174	346	1801
1980	93	231	170	312	2722
1990	49	194	101	205	2430
1995	41	182	97	189	2465
2000[8]	26	167	88	155	2346
2001	24	168	84	170	2428

Source: Statistisk Sentralbyrå. *Folke og Boligtelling*. Skjervøy. Oslo-Kongsvinger (1960-2001)

When I asked local people why Akkarvik's population waned after around 1960, most answered 'I don't know' (*vet ikke)*; very few offered explanations. However, one person suggested, with a mixture of mockery and resentment, that Akkarvik's own elected representatives on the kommune government acted against village interests, when they supported placing the new boarding school *(internat*) in Arnøyhamn. In this person's words, *the politicians who came from here [Akkarvik] wanted them to build the school in Arnøyhamn so it should be easier for kids from different places.* This person denied that the state resettlement program encouraged

[7] While Akkarvik once had a larger population, I frankly doubt but am unable to explain the 1960 population count of 397 people (210 males, 187 females), as listed in the official census.

[8] By 2000, Arnøyhamn replaced Årviksand as the largest community. According local historian Carl Johan Bentsen (personal communication Nov. 2000), the island's actual population as of 1 January 2000 was 409, not the official number of 436. Between 1998 and 2000, the total population of Skjervøy kommune dropped from 2971 to 2932. During the same time period, there were 38 births, 24 deaths, 134 newcomers, and 208 out-migrants in Skjervøy kommune.

Akkarvik people to move, indicating that they left voluntarily, and not always to Skjervøy.

Akkarvik's population declined for the converse reasons as its initial growth. Completion of the post-war reconstruction of North Norway meant that people employed in Akkarvik stores were no longer needed. Fish plants that might have sustained the village failed, as explained in Chapter 3. In any event, between 1960 and my initial fieldwork in 1995, the populations of all Arnøy villages, but most especially those of Akkarvik and Lauksletta, declined. The reasons for depopulation, and efforts to sustain Arnøya's communities are complex, and discussed by locals themselves in the chapters to follow.

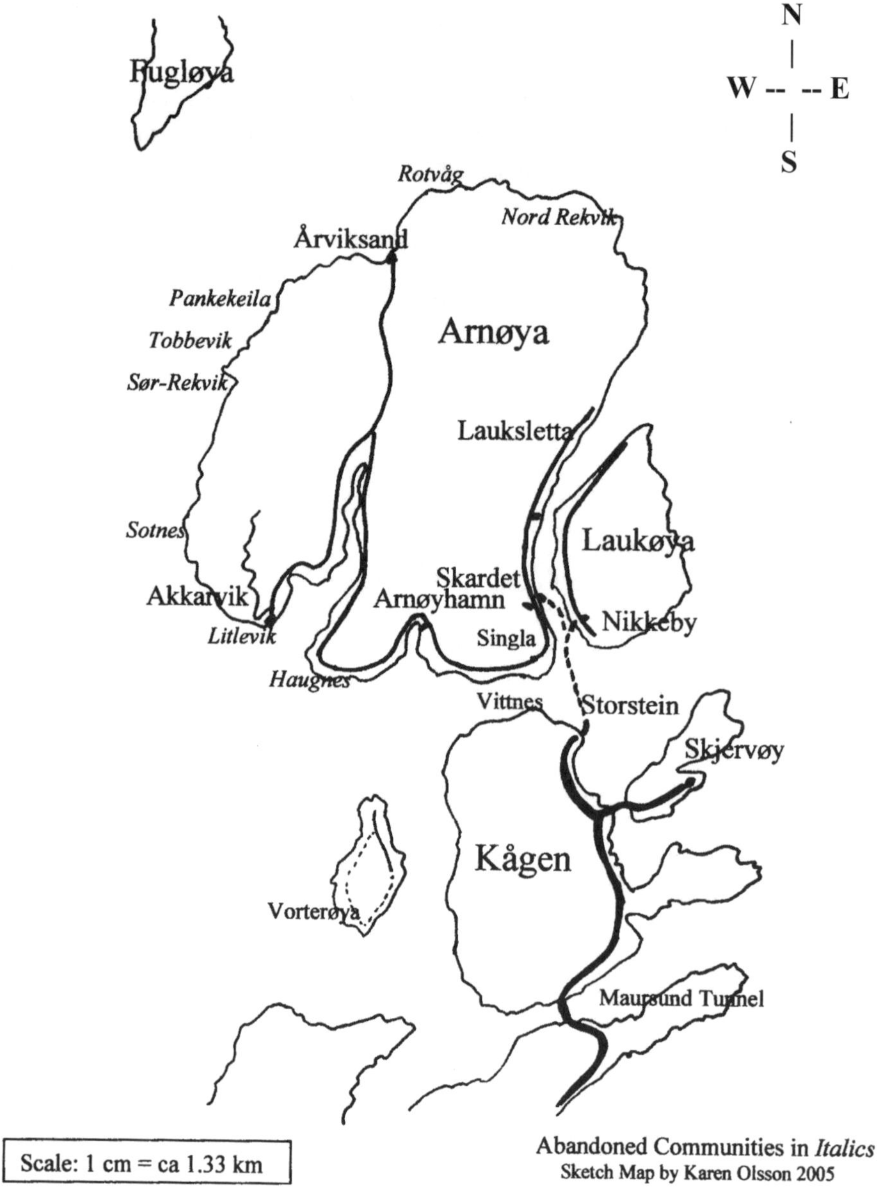

Figure 2 *Arnøya*

CHAPTER 3

TRANSITIONS

Placing boundaries on what is, in people's memories, the seamless tapestry of the past, is always an imperfect and artificial process yet necessary if we are to understand the recent history of Arnøya. The problem of establishing chronologies is even more difficult, as is the issue of cause: *why* things happened as they did. In attempting to address these issues, I have both listened carefully to what people *said* happened and followed the scholarly theory of political economy which emphasizes possible links between the local and the global.[1]

A number of important transitions occurred between the 1960s, a decade of significant change, and 1989, the onset of the fishery crisis. Although an 'isolated' island, some of the changes affecting Arnøya mirror widespread economic changes occurring in North Norway and the world beyond. For example, increased global demands for fresh fish hastened the change to filleting fish in at least one of Arnøy's new fish plants, and this technological change, in turn, triggered changes in other sectors of the local economy and society.[2] Yet as the case of one fish processing company introduced in this chapter (Leinebø and Larsen) shows, the global trend toward production of fish fillets did not affect all Arnøy fish plants or communities uniformly. As elsewhere, local politics coloured the consequences of global change, generating uneven development between Arnøy communities. Like economic changes, political decisions have had, and continue to have significant implications for rural regions, and this necessitates a brief overview of the political system and voluntary associations ordering life on Arnøya. Following this, the transitions discussed below, and others, set the stage for the fishery crisis discussed in Chapter 4.

Local Government and Voluntary Associations

As introduced in Chapter 1, the rural communities or parishes (*bygder*) on Arnøya occupy the remote end of a political continuum stretching to the Norwegian

---

[1] Political economy helps us to understand the changing nature of economy, for example, transformation of the fishery and of fish processing, as well as political changes, both within and beyond Norway. Local histories are assumed to be part of a broader, capitalist context (Roseberry 1988) and, as Eric Wolf (1982) illustrates in his monumental study, local-global linkages can move both ways, with the periphery influencing the center.

[2] Apostle et al (1998: 62ff) provide a good overview of the development of the fresh and frozen white fish trade, which began during the Second World War (See also, Finstad 1999). Arbo and Hersoug (1997) argue that globalization of the white fish trade upset the integrated relationship between fishers, plants, and families in Finnmark.

parliament (*storting*) in Oslo. Between these extremes is Tromsø, capital of Troms Fylke, one of 19 counties (*fylker*) comprising Norway. More immediate still, both literally and in terms of its relevance to local political life is Skjervøy, seat of the municipal government (*kommune*). What interests me most here is local-level politics within the *kommune* itself, and specifically the dynamics between the centre, in this case Skjervøy, and the hinterland or 'districts,' as Arnøy people refer to their island and other rural parts of the *kommune*. To contextualize this dynamic, it is important to understand that Arnøya occupies most of the *kommune* territory but has little of its total population, and even less of its political clout. An Arnøy man quantified the centre and periphery as follows: *Arnøya is [around] 278 square kilometers, Skjervøy is 12 square kilometers. But the main population is living on Skjervøy [island], 2800 and in the districts, we are only about 800 [i.e., Arnøya plus other remote parts of the* ***kommune****]. So, we are a minority. If we get a new mayor [in Skjervøy, from Arnøya], living on Arnøya, we gain much. We've always had revenge between districts and centre. I think you will never get rid of that.*

Skjervøy *kommune* is one of 25 *kommune*s within Troms Fylke. The *kommune* manages various services, such as social, land-use, and others, which are provided to local communities. The *kommune* government has five committees, including one responsible for roads (the technical committee), one for schools (the school and culture committee), one for the fishery (the fishery committee), and so on. Some of these *kommune* committees, such as the fishery committee (*fiskerinemnd*) are linked with relevant administrations at the county (*fylkes*) and national level. When Arnøy people speak of government they are generally referring to the *kommune*, confirming that within Norway's decentralized system of governance, local government is more relevant to peoples' needs than more distant administrations.[3]

A 25-person council, elected by a system one man described as follows, governs Skjervøy kommune: *you have a list of 25 names. If you don't want people, you cross them out, and put in someone say from your own village. This plays community [place] against [one's] party. We vote on a list, but we also vote on the person. The party constructs the lists. The Labour party puts forward a list of its 25 most hopeful, so do the other 4 parties (Centre Party, Christian Democratic Party, Socialist Left Party [note that Norway has 14 political parties; my informant listed those for whom people on Arnøya most frequently vote]).* Although many candidates for the *kommune* council are affiliated with national Norwegian parties, several people suggested that local issues generally transcend national ones during council elections. In the words of one man, *in Skjervøy and rural Norway generally, local issues cause opposing parties to unite around local issues whereas in Oslo, these same parties might remain more distinct. Being a local politician can divide oneself. Your head can be in Oslo, body in Skjervøy, or vice versa. To be a*

[3] Park (1998) presents a good example of the efficacy of kommune government in rural Norway.

right wing politician in Skjervøy is ... only a name to distinguish yourself [from other candidates]... But one must follow some of the principles of the mother party. An Arnøyhamn man explained *kommune* elections this way: *If you really like a candidate, you nominate he/she two times because everyone can make changes on this list. You can give one two votes [i.e., first, second choice] by putting a mark or if you don't like him/her you can cross him out. That is why if the party wants a person you put him two times. Then all that use the list without cutting him out, you have two voices. If every list was unchanged, he would get double the vote of voices. You can vote for 25 altogether.*

Population concentration, essentially to Skjervøy since the 1960s, and more recently, out-migration following the onset of the fishery crisis in 1989, has further eroded the political power of sparsely populated Arnøya. Consequently, Arnøya (and indeed the remaining rural hinterland served by Skjervøy) lags behind the centre in the number of seats elected to the council. This political imbalance accelerated during the 1980s and especially the 1990s, as one man explained.

We [Arnøy people] could have many councillors. At the end of 1980s, we had three councillors from Lauksletta. There was 4-5 from Arnøyhamn and Årviksand. In that period we were [i.e., the council was] more from the districts than from Skjervøy. But today it's the other way around. They are 20-22 from Skjervøy and only 3-4 from districts.

Arnøy's diminished representation on the *kommune* council prompted resolute islanders to invent a more local form of government, the *bygdelag*. The word *bygdelag*, means an informal association of people within a rural community or parish, or cooperation between them.[4] On Arnøya, the word *bygdelag* refers to informal community councils that lobby the kommune government. Lauksletta organized its *bygdelag* in 1981. I asked whether *bygdelaget* were formally recognized by the state? *No, not really. It is something we began in 1981 when we lost our representative on the council in Skjverøy, then we started.* The same man explained the *modus operandi* of the Lauksletta *bygdelag* as follows: *It is very important for us to have this (political organization) because we have no one from Lauksletta on the Skjervøy council. If we want to forward a case to our municipality (in Skjervøy), then we come together, write down our claims, and we feel that we are stronger when all the people of the community stands behind what we are claiming. The Skjervøy council recognizes that local* ***Bygdelaget*** *speak for local people. There are 45 members [i.e., every adult] in the Lauksletta* ***bygdelag****! All the adults.*

---

[4] The word *bygdelag* has several meanings. In one example, not as used here, Frimannslund-Holmsen (1999: 164) emphasizes co-operation between *bygdelag*, when referring to the "surprisingly high degree of co-operation between different rural districts or *bygdelag*," as locals (in his case, an Island off Bergen) called it. For example, farmers from a mountain district rich in trout would exchange trout for barley, which did not ripen in the mountain districts.

During the time period discussed in this chapter, all four Arnøy villages formed *bygdelag*. At Arnøyhamn, pressure by the *bygdelag* on one issue paid off in 1987, as a local man described: *We have an organization called the* ***bygdelag****, we voted for a building for old people [seniors home]. Old people whose children had moved away would have to move off Arnøya. We needed money. We have a federal bank, not private. We got credit from this bank. But our rent couldn't pay our loan, so we made an agreement with local [****kommune****] government to help us. We argued it would be cheaper for government to help us. The [****kommune****] mayor accepted it. The issue was between Arnøyhamn* ***bygdelag*** *and Skjervøy* ***kommune****. The old peoples home was started in 1987 and completed in 1994. There are 6 apartments in it.*

Finally, I note one woman's observation that the ***bygdelag*** *is not political [as in party politics]; it's just the organization working for the place.* In other words, *bygdelag* are local lobby groups, neither organized for nor motivated by partisan politics.

As will be illustrated in Chapter 6, with regard to transportation policies, the localized constituency of *bygdelaget* explains why these informal councils occasionally divide Arnøy villages, pitting one village against another. On the other hand, more 'universal' issues occasionally cause Arnøy's *bygdelag* to unite.

Arnøya also has a number of voluntary associations that meet occasionally to pursue various local goals. One of the oldest of these associations is a local branch of the National Norwegian Women's health organization (*Den Norske Sanitetsforening*), explained to me as follows: *locally, this organization, started just after World War Two; it was needed. They bought equipment that people could borrow, for health, if you were sick you could borrow .... At that time we didn't have electric lights, water closets, and so on, so people had to have things to use in their beds (bed pans). They bought cod liver oil, carrots, and apples to give to school kids in the 1950s. Also in the 1950s, there were few dentists at Skjervøy, so the* ***Norge Sanitetsforening*** *paid for tickets for children to go to Tromsø. The organization also organized public health, for kids to have inoculations.*

Because of the distinctive personalities and histories of each island community, not all local organizations were active in all communities. Thus, an Arnøyhamn man explained the absence of *Den Norsk Sanitetsforening* in his community, as follows: *The* ***Norsk Sanitetsforening****, [is the] women's health organization that raises money for health. We don't have such things in Arnøyhamn; we have different groups than they have in Akkarvik and Lauksletta. We have the* ***idrettslag****, the sports organization, and an active* ***bygdelag****. The ladies meet every Thursday just to talk and make money for equipment for the* ***grendehuset*** *[community hall], and so on. We had things like this (****sanitetsforening****) earlier but I don't think they are very active.*

An Årviksand woman described the organizations of that community, some more active than others. One type of gathering, the community social gathering (*sluttelag)*, reflects a reduction in prohibitions surrounding alcohol consumption, enabled by the gradual wane of Læstadianism, as described in Chapter 1.

Another organization that is quite active in Årvik is the housewives group, *husmorforeninga*, described by one woman as follows: *Not every woman is member. They actually have more money than other organizations. They have done a number of projects, lights for ski trail* (**lysløype**)*, tables for a picnic area, and flowers to elders. They have much money. They raise money through raffles. That's how we use our pocket money. I think they have the house used for church [i.e. the house used for Sunday Læstadian services], they pay the lights and maintenance, and send money to the church, and they provided water to this house. They also send flowers to elders at five-year intervals: 60, 65, 70 and so on. And we used to send flowers to newcomers, did you get one?* [John: No, but that's okay]. I asked one man why is it that each community has different organizations. He answered: *I don't know, it's maybe [different] interests. In another way, the few women who are active in Lauksletta are very active. And the same in Akkarvik. They have a lot of energy.*

Informal associations such as *bydgelag* or *idrettslag*, organized by locals to achieve one or another purpose, are precisely the kind of 'weak ties' enhancing trust between people, and are thus important assets to communities, offering a means by which people cope with rapid or threatening change.[5] Although different associations exist in each community, the number and importance of associations appears more or less evenly distributed between all four communities.

Fish Plants

The 1960s was a time of upheaval (*brytningstid*) in the fisheries, as global demand for frozen white-fleshed fishes replaced that for traditional dried or salted products. Changes in American and European consumer preferences radiated to far off Arnøya, causing the era of smaller boats and larger seine boats (*notbåt*) catching *sei* during summer to gradually give way to larger vessels fishing from centres, mainly for cod. Within a decade or so, this and other changes would lead to an overcapacity in the fishery and to pressure from the state to reduce the numbers of fishers and plants. Between the 1960s and 1995, the number of fish plants on Arnøya dwindled from five to two. In terms that might have been used at the time, North Norwegian fishers abandoned traditional processing methods and adopted modern ones. By doing so, it is perhaps ironic that fishers 'progressed' from producing what was essentially a 'value added' or kitchen ready product (at least in the case of unsalted dried fish), to catching and preparing fresh fish, to be eventually processed or packaged elsewhere.

---

[5] Granovetter's (1985) now classic contention that such seemingly 'weak ties' encourage trust is an important kernel in the currently popular concept of social capital, as discussed in Chapter 5, footnote 2.

Between the end of the war and the advent of Leinebø and Larsen (discussed below) Akkarvik had two fish companies, both operated more or less as cooperatives. People actually called the first of these a Co-operative, one of many co-ops Norwegian fishers established following the war. People also claimed that this co-op went bankrupt although no one was able to recall why. Typical of recollections were those of a Langfjord fisher, who recalled: *they had local owners before Leinebø and Larsen took over. That was not unique to Akkarvik. In many places there were local owners, a cooperative. Akkarvik was the biggest community [on Arnøya] during the 1950s.*

Around 1953, a second company called the Akkarvik Produksjonslag, essentially a fishery cooperative, replaced the previous co-op and was managed by Harry Rikardsen between 1953 and 1957. As part of his duties, Rikardsen also managed the Steamboat office (*Dampskipsekspedisjon*) and telephone exchange.

As said, fishers played an important role in the Produksjonslag, forming a board that decided on prices. The company also operated a store, managed by Johannes Konst. Fish was purchased by the company on notes (*på seddel*), guaranteed by the fishery union (*Norges Råfisklag*), providing fishers credit in the store. Fishers were paid twice a year.

However, the small size of the company meant that it occasionally relied on volunteers to help unload big catches. The company lacked the capital to pay workers overtime (25% more than daily wages) to receive fish at the depot (*fiskemottak*) after 17:00 each day, meaning that some fish lay unprocessed too long, lowering its quality. Inevitably, some fishers began selling their fish to other buyers, who promised 5 or 10 cents (*øre*) more per kilo, and thus not landing their fish at the Produksjonslag. Around 1960, the Produksjonslag, like its predecessor, went bankrupt.[6]

Small, locally owned fish plants also existed in Lauksletta and Arnøyhamn during this same period. An Arnøyhamn man took over his father's plant in that community in 1965 and then listed all *ten* plants then operating within Skjervøy *kommune*. *My father was a fish buyer, before, during, and after the war. I took over in 1965. From 1965 onward, many small fish factories disappeared. There were five on this island; today there are only two. In that period I was one of them. There were two reasons for this change: few fish in the sea and stronger control on quality of fish plants. In Skjervøy* ***kommune*** *there were two plants in Lauksletta [the Nygård family and Skarbruket plants], two on Laukøy at Nikkerby, one here in Arnøyhamn, one in Akkarvik, one in Årviksand, and three in Skjervøy.* Similarly, a Lauksletta man told me about his family's former plant. *Our father had a small fish plant. In earlier times, this place [Lauksletta] had about eight or nine bigger fishing boats. Much fish was landed. There were two fish plants [in Lauksletta]. In the end of the 1970s, both were modernized, and expanded. Skarbruket, for*

[6] I thank Harry Rikardsen's son, of Akkarvik, for details presented in these two paragraphs.

example, before the modernization, had capacity for about 1000 ton, after the modernization, for about 5000 ton. This kind of modernization occurred in all [Norwegian] plants. But then, in the 1970s there was a lack of fish for the plants. Then the government made a plan to reduce the number of plants. They subsidized factories that quit, paying them not to produce. This started about 1980 and lasted until 1988. About two-thirds of all small factories disappeared. Our fish plant was one of these.[7]

Leinebø and Larsen

The story of the firm of Leinebø and Larsen is a crucial ingredient in the recent history of fish plants on Arnøya. The importance of this firm, indeed, for some locals, its infamy, stems from the firm's impatient attitude toward labour, the fact that one partner was raised locally, that each owner championed different Arnøy communities, and because before the crisis beginning in 1989, the company owned fish plants in Akkarvik, Årviksand, and briefly, in Lauksletta. The decisions that the company took regarding which of its fish plants should be favoured were, in retrospect, consequential. The infrastructure of Årviksand would be enhanced; that of Akkarvik diminished. These decisions were based on the two partners' divergent local loyalties, and ironically, on the order in which the partners, and their heirs, died.

Johannes Leinebø, originally from Bergen, and later his son Jan, formed one-half of the partnership while a locally raised man, Aksel Larsen, formed the other half. Aksel and his adopted sister Margit were foster children (*pleiebarn*) raised by Årviksand merchant Amandus Pettersen (who we met in Chapter 2) and his wife Jenny. Adopted siblings Aksel and Margit would later marry each other and have several children, of which only their son Arnt is important here. Leinebø and Larsen replaced the Akkarvik Produksjonslag in the early 1960s and began operating a plant at Årviksand about the same time. Significantly, the plant at Årviksand included a new filleting line and a young woman told me about an older Årviksand woman, now deceased, who was among the first women to make the arduous transition from home and animal barn to the filleting line of the Leinebø and Larsen plant: *[she was] one of the first women who started to work at the fish factory. This was around 1961 or 1962. They earned 2 NOK per hour or so, when she started. She told me what it was like. First of all, 5:00 AM she had to go to her mother-in-law's [house] to milk the cows. First, she puts the kettle on and turns it to number one so it does not boil, just gets hot. Then she goes to milk the cows and her mother-in-law was very strict. If she was five minutes past [late] she was [reprimanded]! She wasn't doing this for cash; it was a family thing. The cows had*

[7] None of these informants referred to the North Norway Plan (*Nord-Norge Plan*) begun in 1952, a Norwegian development policy that used generous loans from a development bank to finance projects intended to reduce regional disparities. However, I suspect some of these locally owned fish plants utilized loans available under the North Norway plan.

May 17 National Day parade, Årviksand 1995. Note Molo in background. (Photo by Karen Olsson)

to be milked. It was her cows as well. She had her laundry hanging over at her mother-in-law's [house] because they had no place to hang it. So she'd bring that home as well. When the clock was about 6:00 she would be getting the children up. She had to be at work at 7:00 AM. There was no kindergarten then. Many times she didn't have time to eat. She did all this. She worked at filleting. And then of course her husband was out in the boat. So of course the women had to do all the work with that too, like washing clothes. And when there were ***akkar*** *[squid] in Akkarvik, they would have to send the people to Akkarvik to pack it. She was one of the first persons to work on the fillet [production line]. I think it was in the early 1960s. They were standing on the floor with fish guts everywhere. There was no ventilation, no air.*

Although Leinebø and Larsen offered wage work and paid fishers in cash, peoples' memories of working for the company are generally not happy ones. One Årviksand man recalled the arrogant manner in which one of the firm's partners treated workers. In his words, *one of [the] Leinebøs [i.e., Johannes or Jan] was here and said a fool thing. 'The women on the fillet (line) don't need a holiday.' Very bad man, stupid.* The same informant continued, *my sister worked here on fillet, for Leinebø and Larsen and she earned 6.5 or 7.0 (NOK) per hour. But the man who owned the factory didn't pay for insurance. So if they got sick they were not covered. The man who owned the factory didn't like it when they organized [unionized].* Another Årviksand man likened the manner in which Leinebø and Larsen treated workers to the historic *Nessekonge* [fish merchants] while another man recalled the antiquated plant where *they used wheel barrels to transport heavy fish around.*

I received conflicting answers to questions concerning the extent the Leinebø and Larsen plant at Årviksand processed filleted fish, as compared to producing traditional stockfish. Typically, one man claimed the people stopped producing stockfish *when we start to produce fillet in Årviksand*, while another man claimed that the firm continued to produce some stockfish. Clearly, however, there was a filleting line at the Årviksand plant and, given increased global pressures toward producing processed fillets, this influenced the fate of Årviksand and Akkarvik, as the latter concentrated on stockfish.

A chain of events and priorities would eventually cause the company to favour its Årviksand plant. An Akkarvik man explained (with understandable loyalties) it this way: *the first Leinebø [Johannes] wanted the big plant to be here in Akkarvik, but [Aksel] Larsen wanted it to be in Årviksand. Leinebø thought that it would be easier to maintain production here in Akkarvik because construction of a* ***molo*** *[such as at Årviksand] would not be necessary. [Next], sometime in the 1970s, the old [Johannes] Leinebø died and his son [Jan] took over, but he didn't know shit about the fishery. They had already built in Årviksand but it was to get bigger. It was Arnt Larsen who ran the plant here in Akkarvik, because the other half, [Jan] Leinebø who took over after his father did not have much interest [in Akkarvik]. The son of Larsen, Arnt wanted the plant to be here, in Akkarvik, but [then] he got cancer and he died. He was in his early 30s. Arnt Larsen wanted them [the fish plants] to work together, in Akkarvik and Årvik, that's what he wanted. Good cooperation between them. He had big plans for this place.*

With young Arnt Larsen's death, Akkarvik lost its sole champion within the company. The same man continued, *Arnt Larsen's mother took over the Larsen share, here in Akkarvik. Margit, the mother of Arnt Larsen, and Jan, youngest son of [Johannes] Leinebø, owned this place [Akkarvik] together, and they go broke in the end. A lot of accidents made things end up the way they did.*

Some locals still *blame* Leinebø and Larsen for going bankrupt, and attribute the company's failure to its chaotic business plan. One local man, who managed the company's fish plant in Årviksand for two years in the early 1980s, put it this way: *they (Larsen and company) went bankrupt because they lacked an idea, a business vision. At that time, the company had little direction. One of the old owners had died and there was uncertainty about what to do. When I was there we got licence for shrimp production but nothing happened. The man in charge [presumably Jan Leinebø] lived in Tromsø. I think his ideas are yesterday's ideas. You know what happened [they went bankrupt].*

In sum, Johannes Leinebø wanted to develop the plant at Akkarvik while Aksel Larsen favoured his adopted Årviksand. Johannes Leinebø died first, leaving Arnt Larsen to manage and promote Akkarvik, even while his father, Aksel, an Årviksand supporter, was still alive. However, local hopes for the future of Akkarvik's fish plant may have died with Arnt Larsen's own premature death, leaving no one to promote Akkarvik. The final two partners, Margit Larsen and the absentee owner Jan Leinebø somehow managed to hold on to both plants and even to add the Skaret plant at Lauksletta before the fishery crisis of 1989 prompted the

May 17th dedication of Second World War Memorial, Årviksand 1995.
(Photo by Karen Olsson)

Norwegian government to reduce the number of fish plants, as described below. As we shall see in Chapter 4, when I discuss the fishery crisis, in the final analysis, Leinebø and Larsen profited from its time on Arnøya.

From Stockfish to Fresh Fish

I consider the post war transition from stockfish (*tørrfisk*) to fresh fish to be very important, especially when combined with the demise of the *fiskarbonde* economy, discussed below. The winter production of dried cod and saithe/pollock (*sei*) was ideally suited to the seasonality of the *fiskarbonde* economy. People caught and hung fish to dry between December and April. However, stockfish production fell victim to the globalization of markets, and as noted, changes in consumer demands toward fresh or frozen white-fleshed fishes such as cod, saithe/pollock, haddock, and others, rather than dried fish.[8] Moreover, Arnøy's extended but inconsistent winter weather, characterized by alternating freezing and thawing, impeded the drying process, reducing the quality of stockfish produced on the island. Consequently, Arnøy stockfish (especially *sei*) was generally shipped to West

[8] Details of this transition are presented in Apostle et al (1998: 63-64, and elsewhere). In addition to this broader context, more locally, people appear to have considered the production of stockfish as anachronistic, and inimical to modernization. Contrary to my own view on the matter, more than half (54 percent) of the Arnøy people I interviewed in my adult questionnaire either *agreed* or *strongly agreed* that the transition from stockfish to fresh fish had been good for the economic development of their communities.

Africa, as opposed to the more lucrative southern European markets. One Arnøy fisher explained his stockfish enterprise, what happened to stockfish markets, and, eventually, to small boat fishers. *I dried it and sold it to the* ***Nessekonge*** *in Havnnes [laughing]. The dry fish king in Havnnes. We were too much dependent on a few countries in Africa that imported the dry fish. And the Norwegian state helped them import with its development money to give them food. They used development money to give food to some poor African countries. That stopped. The market collapsed for* ***tørrfisk****. Nigeria didn't buy it. War -- the Biafra crisis, you have no markets for* ***tørrfisk*** *during the Biafra crisis. They weren't allowed to buy fish. War and blockades.* [John: when you stopped producing dry fish, what did you do with the fish you used to dry?] *I sold them. I delivered everything to the fish factory. When you delivered it to the fillet factory you have to take more care of the quality of the fish. I earned about the same. I had less work with the fresh fish than with the* ***tørrfisk****.* [John: when did most stop hanging fish?] *The end of the 1970s. The smaller boats they had a special problem. The smallest boats, those [mainly older] people rowed, they'd hanged fish all the fall. When they couldn't dry it, they had no place to sell it. [When this happened] they couldn't fish.*

An Arnøyhamn man explained the disappearance of Nigerian markets and how this reverberated back to Arnøy fish plants. *The Nigerian market for stockfish disappeared about 1970. Then we had to change over to another method of processing fish and freezing was too expensive for all existing plants. So only a few survived.* The mayor of Troms Fylke in Tromsø discussed stockfish markets and then responded to my question of why North Norwegian salt fish (*kleppfisk*) continues to be commercially dried in West Norway. I then asked why stockfish -- so much a part of North Norway's past -- could not be part of its future? The mayor responded: *stockfish is still alive as a product. But it's a difficult product. High quality stockfish, dried in Northern Norway after the season of frost, has the market in Italy and other Catholic lands, and to a very high price. It's a very profitable niche, but it's only a niche. The stockfish with some volume,* ***sei****, was meant for the African market, but political and economic problems in those countries, more and more destroyed that market. And today, it's very difficult to sell dried* ***sei****, haddock and that kind of product. It is an interesting [additional] question: why do we not dry our salt fish, here in Northern Norway? Traditionally, it was almost impossible, because the fish were dried on the rocks, but today, it is artificially dried. There is no technical reason why we can't dry fish here, but the companies who tried it so far have not succeeded. The value adding still takes place in Western Part of Norway.*

Consequently, wet salted cod continues to leave Arnøya for drying in Western Norway. Meanwhile, wooden fish drying scaffolds fall into disrepair, haunting reminders of a production method now replaced by fresh fish plants. And finally, employment on new fish filleting lines attracted people away from their sheep barns and small boats, both gendered practices integral to the old *fiskarbonde* economy.

Demise of the Fiskarbonde Economy

This section describes the demise of the *fiskarbonde* way of life, but it is important to note that some vestiges of it survived during the 1995 fieldwork. Each community had at least one sheep farm; two were mixed fisher-farmer operations, and two were full-time sheep farms. However, by my return in 1997, the Akkarvik farmer had ceased farming and other farmers described new challenges. Without exception, sheep farmers claimed that declining state subsidies were undermining farming. In the words of one farmer, *these subsidies included money for the number of animals we have, for the number we let go onto the landscape, for the cultivation of land, and for the maintenance of agriculture in the districts. The price for lambs has not increased since 1982, and but for the subsidies, we would make nothing.* Farmers blamed declining subsidies on Norway's new relationship with the free market policies of the EU, through the EEA, as mentioned in Chapter 1. Thus, during the years since my fieldwork, European market policies challenge both fishers (Apostle et al 1998: 102) and the few remaining farmers.

Placing temporal boundaries on the end of Arnøy's ancient *fiskarbonde* (fisher-farmer) adaptation is difficult. The Second World War fundamentally altered where people lived, especially after the 1944 deportation. Because the war created something of a boom in the local economy, the post war years unquestionably made people rethink the relationship between work and society. Unsteady post-war markets for stockfish, coupled with the new demand for fresh fish also played a role insofar as the seasonal rhythm of fisher-farmers was well suited to the winter fishery and to the drying of stockfish. On the other hand, the year round filleting of fish not only required full-time fishers and larger vessels to catch continuous supplies of fish, but, back at home, filleting lines attracted people (often women) from the barn and field to the factory floor. My Arnøy data suggests that the older adaptation remained the primary household economy on the island until the 1960s.[9] Yet I do not wish to romanticize the older pattern, which one Arnøhamn man dismissed as *survival farming*. Indeed, one of the reasons for the demise of the *fiskarbonde* economy, with its small-scale fishery component, was that until the 1960s, there was no viable alternative to it. The *fiskarbonde* economy was, as a former Newfoundland fishery union leader (perhaps too callously) once characterized the fishery, the employer of last resort.[10]

---

[9] Apostle et al (1998: 74), citing Drivenes (1982), observe that Brox may have "inflated the importance" of the *fiskarbonde* economy during the post war period; my data agrees with Brox, that the fisher-farmer adaptation survived into the 1960s.

[10] This frank, but politically incorrect, opinion was expressed in 1993 by Richard Cashin, former President of the Newfoundland Food, Fish, and Allied Workers Union.

Viewed another way, one might argue that occupational pluralism did not die with the *fiskarbonde* economy, it merely shifted track. Thus, whereas in the older economy women had primary responsibility over home and barn, while men fished, often far from their homes, nowadays women work in the fish plant, school or social services, while men fish or work at the plant.[11] One final point helps situate the claim of some people that the death of the *fiskarbonde* economy occurred because it denied people summer holidays! As explained in Chapter 5, nowadays the fishery and fish plants close for the month of July; some locals fly south to the Mediterranean or Azores for holidays. The local interpretation that the older, pluralistic economy died because it provided no holidays is, for me, not convincing. Having said that, I begin with it.

One man explained the demise of the *fiskarbonde* economy as follows: *one of the reasons why they stopped the combined farming was that they earned enough so that they could take a vacation.* Another man made a similar point, and advances our understanding of changing expectations of consumption and of space. *One of the reasons why the small farms disappeared was that in the old times, nobody ever left their homes in summer for a holiday. The small farmers did their harvesting in summer and was having their sun bath while they harvested! But in these new times, everyone should have a car, leave their place in summer and then you didn't want the cow holding you back. That old way of living is historical; it doesn't make sense today.*

A more likely explanation for the demise of the *fiskarbonde* economy concerns increased priorities now placed on education, and youth out-migration (See Chapter 7). One man opined: *the reason was that everyone had a chance to get an education, so they left, got the education, and didn't come back.* A Langfjord man put it this way; *it started in the 1960s and went on to the 1970s. I don't think this [demise of the fiskarbonde economy] made people go away. The main thing was education. There came a time when you had to take an education. Everyone said to you 'go to school.' The children had to go off the island and into Tromsø. A few came back.* An Akkarvik man hinted that the peoples' expectations of predictable salaries and work schedules were changing and that this was why the older economy died out. *People wanted to get paid, to pick up a salary. When I was fishing, I had to use my vacation in summer to make hay. I had free time because I was working for myself.* A Lauksletta man added an historical dimension, stressing new demands by fish plants for labour, and new consumer expectations. He said, *to go to the historical thing, this place [Lauksletta] is about eight kilometres [along the shore] of land and you have small farms, with fields around and that was the structure of this place. People have their own; they were self-served [sufficient]. And in the summer people were doing their harvesting on the grass and in the*

[11] Scholars including Blekesaune (1991), and Blekesaune et al (1993) monitor changes in the gendered division of labour of Norwegian farming, sometimes characterizing it as 'pluriactivity' (Hetland 1986). Another study examines the professionalization of women in Norwegian farming (Haugen 1990).

winter were hired on the fishing boats or fish factories. This combined lifestyle is no more. During the 1960s, this changed. People asked for full-time jobs, and you got big fish factories with the aim of having full time work for the people. But that didn't work. Because the government, in the 1960s, if you want to build a factory, the government always asked, 'how many employees will you have?' That was the main question. This led to overcapacity. It was that the policy of government was to make as many as possible full time jobs.

Some people stressed the growing need for money, especially given the inflationary consequences of the development of the Norwegian petroleum industry, in the 1960s. An Arnøyhamn man described the impact of oil, not only on the economy but on widespread perceptions that combined fishing-farming was an anachronism. *I used to say, the first one who found oil in the North Sea, he should have a nail put in the hole and then should have said nothing, to anyone. In Norway, it went much better before the oil than it is today. Today we are totally dependent on oil; the whole economy is driven by oil. I think that if it had not been for the oil industry, fishing and farming would have been valued in a much better way in the [public] opinion than now. Oil is the future.* Similarly, a Langfjord fisherman blamed increased local requirements for cash on North Sea oil, as follows: *the income level increased so much in fishing and factory work that the farming work hung after [lagged]. It became too much to have sheep compared with bigger things. The bigger coast boats were fishing the whole year around. It was a natural development.* [John: why did the prices increase during this time?]. *Now we're in national policies. You have inflation during that time. In the 1970s the oil industry pressed the income and prices extremely fast.* The same man explained how efficient new fishing technologies, and high fish prices, combined to enable men like himself to become petty capitalist fishers and how the improved income from year round fishing made part-time farming unnecessary. I asked when did it become economically advantageous to fish all year? *When you got better and more effective equipment, between 1963-70. [Fishers] turned over from the smaller boats to bigger, coast boats, with better comfort. The nets turned from hemp to nylon. This started in the early 1960s and developed during the 1960s. That was a very special thing. More effective. We got equipment to pull the long-line and also the net from the sea, by hydraulic wheels. You didn't need arm force to pull the nets. You can have more nets out at the same time, than before. There was also the mechanical* ***jukse*** *[jigger].* Another man suggested that the wane of traditional economic pluralism led to increased specialization of either fishing or agriculture. *[People] started to do just the fisheries, they got better paid, and thus didn't need to keep a cow. And the ones, which kept on with the agriculture, they got more. Instead of one cow, they had several (like an Akkarvik woman), who had her last cows in 1988 -- and these were the last cows in Skjervøy Kommune). So they could earn money from it, instead of just having one.*

Some people lament the quality of life lost to the new economy. As one man put it: *we didn't make a lot of money but we had a good quality of life. [We had] satisfaction of the work.* Yet I asked one woman whether her move from barn to fish plant was difficult, in other words, did she miss farming? *I started to work in Skaret [Lauksletta fish plant] before stopping having sheep. So I didn't miss the sheep.* [John: So the new job meant that you didn't need the sheep anymore?] *No doubt about that.* Another woman made a similar point. *When the old ones stopped [having farms], the young ones didn't follow; they didn't need to do it.* [John: is it right to say that people only practised the *fiskerbonde* lifestyle because they had no other choice?] *Yes. I had to do it to live but my children did not need to. Take my daughters-in-law, they didn't want to farm.*

Numerous economic and cultural changes occurring during the 1960s, including modernization of the fishery, use of larger vessels, opening of new fish plants, and improved family incomes, would all led to the demise of the *fiskarbonde* way of life. As one fisherman put it, it was *an economic matter: if you fished all year, the income was better.* Soon, dazzling television images from afar would present a world totally unlike the cold realities of seasonal work with sheep or fishing. New roads led elsewhere and local youths took them.

By the 1980s, many of these transitions - fish plants, filleting lines, schools, roads, and others - were unevenly distributed between Arnøy's four communities. This differentiation of economic and institutional capital would affect each community's ability to respond to the fishing crisis of 1989. Moreover, the 1980s would be a decade of economic growth and easy credit, locally referred to as *Yappetid.* Generous credit would eventually lead to a crisis in the banking system. Coupled with a severe decline in the availability of cod, these events signalled the onset of the fishery crisis, to which we now turn.

CHAPTER 4

CRISES ON THE SEA AND ON THE LAND

This chapter examines some of the main causes and consequences of the short-lived Norwegian fishery crisis that began in 1989. As in other economies primarily dependent upon a finite inventory of natural resources, fishing economies occasionally experience resource shortages; other fishery crises preceded that of the late 1980s, and others will follow it. However, I contend that fishery crisis beginning in 1989 was more than simply a scarcity of ground fish and that its impact was compounded by a financial crisis on the land. Just as there were several causes, so there were many consequences. These include a serious erosion of economic capital, such as for example, the loss of large fishing vessels and of fish plants. These and other consequences fuelled out-migration, reducing the populations of all Arnøy communities.

In what follows, I first examine the two main causes of the crisis, as suggested by the chapter title. I then turn to some of the consequences, focussing on the loss of economic capital. Later on, Chapter 7 examines another consequence, youth out-migration, which continues to weaken the viability of two of the island's four small communities. But for now, let us look at the fishery crisis.

Natural causes of the 1989 crisis include invasions of seals,[1] especially on the Finnmark coast, shortages of caplin (*lodde*), and consequently, of cod (*torsk*). Economically, the end of what locals call *yappetid*,[2] a period of easy credit and over-spending, and consolidation of Norwegian banks, led to foreclosure of fishing boats and other capital equipment, and to considerable 'restructuring' of the local economy. As the demographic data in the Table 2 in the previous chapter suggest, whole families left Arnøya. Although less tangible, the economic and cultural precursors of the fishery crisis may be inferred from informants' transcripts, specifically, accounts describing the difficulties fishing skippers had recruiting reliable young crews during the 1980s. By this time, young men no longer considered work on a fishing boat as appealing as their fathers had. This problem, as much cultural as economic, outlived the crisis, suggesting that Arnøy youth

---

[1] Eikeland (1993) traces the history and impact of harp seal invasions on the fisheries of three fishing communities in Finnmark.

[2] In Winter 2003, Ms. Monica Playfoot, Norwegian and my student research assistant at Memorial University, uncovered the etymology of the Norsk word Yappatid, a word not found in the Norwegian dictionaries I consulted. The word is a corruption of the English acronym YAP meaning 'young aspiring professionals' and used in Norway since the mid 1980s to refer to persons whose attitude and behaviour suggest a professional career orientation. Presumably, Hansen's (1997: 236) phrase 'yuppie years' refers to yappatid.

increasingly want careers ashore, with ominous implications for the fishery of the future.

Try as I did to encourage informants to explain the etiology of the 1989 crisis, I most often heard accounts of particular and sometimes conflicting details associated with the crisis or, less frequently, I accepted condolences for the sorry state of the Atlantic Canadian fishery. Several people claimed that the 1989 crisis had largely spared Skjervøy *kommune*, or that its effects were more catastrophic elsewhere, especially in Finnmark (these claims are largely true).[3] Yet there were local repercussions: a dramatic loss in the number of fishing vessels, of people, and of living standards: the crisis affected Arnøya.[4]

I order my presentation of the crisis by first presenting narratives describing natural causes, where the crisis appears to have begun. Once fishing catches and incomes dropped, it became increasingly difficult for families and businesses to repay outstanding loans. I then turn to the banking crisis of the late 1980s, arguing that it limited the ability of fishers to withstand poor catches. The remainder of the chapter discusses two such impacts of the combined fishery and banking crises: transfers in the ownership of fish plants, and the loss of Arnøy's large fishing vessels.

Crisis on the Sea: Seals, Caplin, and Cod

One Arnøy fisher's explanation of natural causes behind the crisis, began with harp (Greenland) seal invasions, first in Finnmark, and later, further south. He explained: *[in Finnmark] the cod moved out of the fjords when the seals came in. The seal invasion stopped around 1988 or 1989. The cod then went back into the fjords. The seals worked south, arriving here [Arnøya] in 1986 and 1987. In 1987, the cod were down to 500 metres, because the seals scared them down. It was very heavy to haul up nets. It took more expensive nets, thicker ropes, and so on. When the seal invasion started, the [fishery] crisis started too, 1985 or 1986. There was a seal invasion here; thousands of seals, Greenland seals. We made more income on seals than on fish. The seals destroyed the nets. You got seals down to 400 metres. You didn't catch any cod, because they tried to get away from the seals.*

Many people emphasized how commercial over fishing for caplin *(lodde)* starved cod. Representative of comments about caplin is one fisher's answer to my query as to what happened to caplin. He said, *big boats fishing it. [The caplin*

---

[3] General overviews of the Norwegian fisheries crises of 1989-90 may be found in Jentoft (1993) and Apostle et al,(1998: 274-75). Pettersen (1996) focusses on the coping strategies used by Lofoten Island householders during and after the crises, Gerrard (1995) on how the crisis mobilized women's leadership roles in Finnmark, and Pettersen (2000) on how professionalization of the fishery following the crisis forced women out of the industry.

[4] In my adult questionnaire, the most common impact of the crisis on those questioned was lost income (64 percent), followed by loss of job (18 percent). Respondents cited out-migration (22 percent) as the most important consequence for communities.

fishery was] stopped in 1992. The big boats are fishing with ***not*** *(seine), Spanish, Icelandic, through storms, nothing could stop them.*[5] An Akkarvik fisher explained that because fishers *were taking so much lodde, the seals had to go for the cod.* Another fisher explained one detail of local knowledge customarily used to evaluate the health of cod stocks, and apparent during the crisis. Small and discoloured cod livers are an indication of the scarcity of food on which cod depend and, by extension, of the health of cod stocks. In his words, *when cod livers are black, which we call* ***svartlever****, they are hungry. When fish have plenty of food that means that their livers are big and white. [Moreover -- and perhaps obviously] the cod had little food in them or we can see that the fish are standing up [remaining stationary], and we know that the long line gets much fish; this means they are hungry.* People also remembered that cod were thin and when cod were boiled their flesh flakes separated as if there was little to hold the fish together.

Crisis on the Land: Oil Prices and Banking

Although not mentioned in studies on the Norwegian fishery crisis of which I am aware, my informants claimed that the fishery crisis coincided with and was exacerbated by a financial crisis ashore, specifically, the restructuring and consolidation of Norway's commercial banks. My informants consistently alluded to this financial crisis, stressing its impact on the fishery crisis. Most tied that fishery crisis to *yappetid,* so I begin with this term.

The 1980s was a decade of economic growth and easy credit, referred to by Arnøy people as y*appetid*, a term explained above in Footnote 2. Although not all informants knew or used the word, and some spelled it with a 'j', y*appetid* was generally associated with an era of easy credit, which would eventually lead to a crisis in the national banking system, and in turn, would exacerbate the economic consequences of the fishery crisis. Typical of local explanations of y*appetid* was that of one man, who said: *in the early 1980s, the banks offered easy credit: yappetid. The bank crisis was controlled by government and politicians, who wanted to get control over unemployment.* [John: what was the bank crisis?]. *The banks never asked how much you wanted, they just loaned. The government had to use a lot of money to support these banks. They also had to get money for all these people who suffered from this bank crisis. This occurred between 1985-1990. The government had to pay all this money into the banks in 1990 and 1991.* Another man claimed that Arnøy people like him were less prone to borrow than people in larger centres. I asked about local borrowing: *not so much here, in the small places, where people have been taught by their fathers and mothers; 'buy when you have the money, don't borrow so much.' But in the towns, people taking small loans, 100,000 NOK, buy a motorcycle, a car... Nobody asked how shall you pay it back.*

[5] A majority of people (62 percent) in my adult questionnaire attributed over fishing as the cause of the 1989 fishery crisis. The second most common cause (cited by 22 percent of respondents) was a shortage of caplin, in turn caused by over fishing.

People today want the best, so fast. They don't wait. I hate to give the bank a lot of money they don't give me. I borrow 500,000 NOK and give them 300,000 NOK in interest. I had a mortgage of 180,000 NOK on this house. I built it in 9 or 10 months. I was working in Årvik during the week, I came back Thursday evening. I worked on the house Friday, Saturday, and sometimes Sunday. [Today] I have no loans on this house.

The *yappetid* ended in 1989 or 1990 with a crash in the banking system that in turn led to a restructuring and consolidation of banks. In 1988, local branches of the Sparbank [the 'savings bank'] closed at Lauksletta, Arnøyhamn and Årivksand, throwing several people out of work and arousing considerable local resentment. The end of the *yappetid* also led banks to repossess fishing boats and, in other cases, caused boat owners to sell their boats so as to avoid the humiliation of repossession by the bank. Coupled with a severe decline in the availability of cod, these events signalled the onset of the fishery crisis.

Although the *yappetid* of the 1980s precipitated the fishery crisis, what ultimately lay behind this period of easy credit and overspending? A 1991 article in *The Economist* explains some of the background of Norwegian banking difficulties during the 1980s. Norway's banking woes began with *'bad [economic] policies'. Flush with cash in the early 1980s, Norwegians... borrowed heavily to take advantage of generous tax breaks on interest payments. The resultant spending spree lasted until 1987, when the government tightened fiscal policy... [Then] in 1986 oil prices fell below $ 9 a barrel... In 1987-89 over four-fifth of Norwegian banks' losses were due to dud corporate loans... After huge losses in 1988 and 1989... [many Norwegian banks saw] mergers as one solution.* (*The Economist 1991, 73*).

It appears then that financial troubles on the land, ultimately caused by government fiscal policies encouraging spending, and a sudden drop in oil prices, intensified problems already occurring in the sea. Although beyond the scope of this case study, one wonders about the consequences of removing one or another element in this chain of causation, such as oil prices. Might the banking system have survived if oil prices had remained stable or even increased, and how might this have trickled down to efforts by local fishers to withstand low catches of cod?

As matters developed, once fish stocks dropped, fishers expanded their fishing effort so as to attempt make loan payments. Unquestionably however, Arnøy people saw the *two* crises as interrelated, as one Årviksand businessman explained: *we came through the [fishery] crisis but we had to work hard. It was both a bank and fishery crisis. The land businesses managed to recover more quickly than fishers. The two are interrelated.* Another Arnøy businessman described how the dual crises affected Norway's export economy. In his opinion, *the bank crisis made trouble for more people than did the fishery crisis. But for the country [Norway], the fishing crisis was harder. Norway lost a lot of customers around the world. We had to start from the beginning building new markets.*

As noted above, an immediate consequence of the consolidation of banks on Arnøya was the closure of the Sparbanken branches in Lauksletta, Arnøyhamn and Årviksand. The Årviksand branch closed in August 1988. A man who had originally worked for Sparbanken in Tromsø, where he met and married an Årviksand woman, and then moved to Årviksand to open a branch of the bank there recalled, *I came here in 1979 and operated 9 and one half years afterward, until 1988. I opened the bank here. This was Sparbanken. Business was not bad as there were five big fishing boats [See below] here in Årviksand. There was more activity. They [fishers] put money in the bank after each trip. I think the bank made good money here.* ***Yappetid*** *- every one could go to the bank and get money, to buy new radar for the boat, a car, or whatever. Easy money. Not everyone could pay back. I know everybody here, so I know how much money they have. So, I knew how much to loan. But in Skjervøy or Tromsø, they loaned to everybody. It [**yappetid**] began in 1983 and lasted until 1987 or 1988. In 1988 Sparbanken Nord went bankrupt. Then the Norwegian government loaned Sparbanken Nord money, then Sparbanken Nord and Tromsø Sparbanken got together [merged] and in 1991, the combined banks began to make money. During this time (about 1987) several of these [local fishing] boats were sold, first Viggo Bakke [the vessel], Årviksand was next. The Årviksand [vessel] didn't have problems with money but it was difficult to get people on board the boat. Young people wouldn't work on board the boat, they chose to go to school. After they sold the boats and young people began moving away. The Beverly Jane sank at Barents Havet (Barents Sea).* The same man went on to discuss the closing of the Sparebank in Årviksand. *The bank closed in 1988. Everybody was angry. Frank was chairman of the local* ***bygdelag****; they protested, unsuccessfully. They wrote letters to the papers. And the mayor [of Skjervøy] travelled to Tromsø to the bank's head office. We were two working in the bank, the other was Rolf's wife, she lives in Skjervøy now. I wrote to my head office in Nordreisa, suggesting we only work one day a week and close the offices in Lauksletta and Arnøyhamn, one day a week. The bank manager in Nordreisa said it was okay with him, especially given that the Årviksand bank had 30 million NOK in deposit. But officials in Tromsø overruled him. People here were angry and put their money in the Postbanken [the post office bank]. I started work in the same bank in Skjervøy, commuting. I continued that commute until February 1991. I had no loan [mortgage] on this house, and the car and I didn't need to make so much money. That's when I went on my own with accounting and I also have a fishing quota for 3000 kilos.*

Once money got tight, banks called in outstanding loans on boats, houses, and other property. In the words of one Akkarvik fisher, *the bank crisis had a greater influence on coastal communities than the crisis in the fisheries.* Some families responded to the dual crises by leaving Arnøya. Overall though, the island may have unwittingly benefited from sluggish economic conditions elsewhere in Norway. As one man astutely observed, *the bank crisis and the end of* ***yappetid*** *came at the same time. The fish crisis came at the same time. People here with nothing to do, had no reason to move, as the whole country was affected by these crises. There was nothing to leave for. If there had been no unemployment in the south of Norway, and the crisis here, it would have been different. What saved us was the unemployment elsewhere.*

In short, although the interrelated crises on the land and in the sea were not as severe as elsewhere in northern Norway, such as in Finnmark, the crises on the land and sea had consequences for Arnøya, to which I now turn.

Consequences of the Crises: Loss of the Fishing Fleet

The robust economic times implied by the word *yappetid* in the 1980s encouraged several Arnøy fishers to invest in fishing vessels over 50 feet. There were at least five of these large vessels [although medium sized when compared to those of the Ålesund fleet in southwest Norway] in Årviksand during the 1980s, and others were purchased elsewhere on Arnøya. The five vessels in Årviksand include two (one person said three) named the Årviksand, the Beverly Jane (originally from Canada), the Viggo Bakke (the second incarnation: see below), the Kartøy, and the Alfredson. In what follows, I discuss three of the five in some detail, most especially the cases of the Viggo Bakke and the Alfredson. These three cases illustrate some of the problems the dual fishery-banking crises caused for the coastal fleet.

It is important to emphasize that each of these five boats employed several fishers (normally four to eight) and that fish landed created work for others, often family members, in fish plants. Put differently, each fishing vessel greatly enhanced the economic capital of its homeport, both by employing fishers and plant workers. During the 1980s then, these five and other smaller vessels supplied fish plants such as those owned by Leinebø and Larsen. I repeatedly heard how large vessels and fish plants are interdependent, supporting both their own and other communities.[6] But as the 1980s played out, declining catches, and outstanding loans, would eventually claim four of these five fishing vessels. It may also be noted that informants' memories sometimes clashed, often in irreconcilable ways, in some instances because the person or persons who knew precise details no longer lived on Arnøya and whose voices were, consequently, silent. This 'methodological' problem surfaces in the following two accounts of the vessel Viggo Bakke.

One person's account of the Viggo Bakke begins in 1949-50 when an Årviksand man, whom I shall call Tor, purchased a 52-foot boat built at the Arendal shipyard in southern Norway. Tor named the vessel Viggo Bakke after his son, Viggo, with whom he owned the boat and who also fished from it until they sold it in the spring of 1980. In 1979 or 1980, father Tor, Viggo, and his brothers Ole and Erik purchased a new boat, built at the Rognan shipyard in Nordland Fylke, which

[6] For a discussion of this interdependence (or embeddedness; See Apostle et al, 1998) and Jentoft (1984: 151ff).

they also christened the Viggo Bakke. The second Viggo Bakke was constructed of wood, with an aluminium over-deck, and cost of around 1.7 million NOK, excluding eventual fishing gear and rigging. Tor and his three sons began fishing in the summer of 1980 and added a younger brother to the crew in 1981. The vessel enjoyed two or three good years. However, perhaps after father Tor pulled out in 1983, brotherly antagonism over leadership increasingly undermined the fraternal venture. One brother was supposed to be skipper but a bleeding ulcer often kept him ashore; another brother pulled out in the summer of 1983. The two remaining brothers attempted to continue but encountered difficulties recruiting local labour, as the few locally available crewmen chose to work on more stable vessels, where they could earn more money. Next, in 1985-86, the crew of the Alfredson (see below) leased the Viggo Bakke while the Alfredson was being renovated. Following this, the two brothers who owned the Viggo Bakke rejected an offer to sell the boat in the summer of 1986, but, increasingly plagued by debts, the two finally sold it just before Christmas 1986, to a buyer in Vesterøy.

The youngest brother's account differs slightly from the above account, directly attributing the sale of the Viggo Bakke to the fishery crisis. He recalled, *I started with the new Viggo Bakke in 1981. It was brand new, 57 feet long. My father was aboard for 2 years, but he stopped in 1983 and bought a smaller boat. My brothers [then] owned the Viggo Bakke. My brothers, Viggo, Ole, and Erik bought it; I was a [hired] man aboard it. The four of us fished. In the summer of 1986 we were fishing with the* ***reke*** *[shrimp] trawl. After the summer the problems began. We didn't get enough* ***reke****. We had bad weather and this prevented us from fishing. We were fishing in Finnmark and the Barents Sea. We began on* ***torsk*** *here, but there wasn't enough* ***torsk****. We fished on Fugløya bank. We fished outside Lopphavet, with long line. Then in 1987, we were fishing 30 tons of* ***torsk****, it was not so much money. And the banks they wanted the money. This was Sparbanken. We got a letter, saying that the bank must pay by such and such a date. So, we sold the boat, because they wouldn't like it that the bank would not take it.* [John: Why did you sell first?] *It's not so good to lose the boat, 'you're not so professional a fisherman.' After we sold the Viggo Bakke, I worked on a boat called the Årviksand. And I was on board for one year; then it was sold. That was in 1988. After that I worked on the Alfredson. I was on board it until 1990. The bank took the Alfredson in March of 1990. The boat was to go to Tromsø and the bank would sell it there. But the people here and in Skjervøy kommune bought it back. We needed it here, the plant needed fish, which for a time we had to buy from other boats.*

The sale of the Viggo Bakke prior to its probable foreclosure caused all except one of the four brothers to leave Årviksand. The brother who remains recalled how his brothers acquired careers outside the fishery, in other parts of Norway. *Viggo left for Skjervøy when the bank closed here [his wife worked for the bank]. They (my brothers) sold the boat in 1987. Erik, who had been skipper, he moved to Oslo and took an education. Today, he's a house builder and farmer in the interior of southern Norway. Ole, he took an education [at the state fisheries school in Vardø], and today he's a fisheries controller [for the Fiskeridirektoratets Kontrollverk], on the coast guard. He lives in Vardø.*

In the case of the two different vessels called by the name 'Årviksand,' the same informant continued, *the case with the Viggo Bakke and Årviksand was not so good. There were two [vessels named] Årviksand. The first Årviksand was the same as the Viggo Bakke. The second Årviksand was like the [vessel] Alfredson. [Brothers] Gunnar and Roar bought the second Årviksand. Gunnar bought it around 1982; it was a new boat. The last year Gunnar had this boat he had many young people working for him. The first year this arrangement worked well but after that the young people came and went [i.e. there was an unstable labour force]. The last year Gunnar had the boat I was on board and there were also older fishermen, but it was too hard for them. So Gunnar sold in 1988 because it was too much work for too few returns.*

The 52 -foot Kartøy would also be sold. Indeed, when I began my fieldwork in 1995, only one of the five large vessels operating out of Årviksand a decade earlier, the Alfredson, remained. Yet the Alfredson's voyage would also be stormy. The Alfredson would go bankrupt at least two times but like the proverbial cat with nine lives, would, at least during my time on Arnøya,[7] find its way back to Årviksand. The story of the Alfredson is complex, intertwined with local, regional development and national fishery policies, and was difficult to compile. What follows is more a summary than the complete saga.

Like the histories of most boats on Arnøya, that of the Viggo Bakke and Alfredson begins with a previous boat or boats. In 1945, a local fisher purchased a vessel called the Terningen, which he sold to a Finnmark fisher in 1958. The next year, he purchased the Arnøybuen, and sold it in 1970. This fisher and his sons purchased the brand new Alfredson in 1971. He fished with his sons aboard the Alfredson for one year and then, in 1972, he purchased a smaller fishing boat, a *sjark* called Ann Karin, and fished on his own while his sons (first three, later two) continued to own and to compose the nucleus of the Alfredson's fishing crew for the next quarter century. Additional crewmembers were drawn from the village of Årviksand and like the brothers, most worked regularly on the vessel for many years. By 1995, one crewman had been working on the Alfredson since in 1971, another since 1973, and a third since 1985.

During most of its career, the Alfredson, like other vessels, fished for cod, herring, prawns, and other seasonally available species. A local fisher who fished on the Alfredson during the early 1980s remembered, *when I fished with the Alfredson in 1982 we didn't go far from here. We fished close to here. There were not so much torsk but good with prawns.*

---

[7] I learned upon returning to Arnøya in November 2000 that the vessel had been sold, but by 2003, was again owned in and fishing from Årviksand.

In the mid-1980s, the Alfredson's fortunes changed in two important ways. In 1985, with only about 600,000 NOK still owing on the boat, the brothers decided to add about two metres to the Alfredson's stern, a renovation which cost about 4.2 million NOK, and was made to enable use of a stern trawl, as explained below. One Arnøy man, speaking with the precision of hindsight, placed a moral hue on the renovation: *[The brothers] took out a loan to make the Alfredson longer and build a deck. He [the skipper] asked the bank and they gave him all the money he needed. It was a bit of* ***yappetid****! The money was easy to get. But the man on the street wouldn't have been so lucky.*

Secondly, around the same time (1985 was the date I heard), Ministry of Fisheries officials in Oslo awarded one of five North Norwegian trawl licences to the Alfredson (others were awarded to vessels fishing from Karlsøy, Troms country, and to Nordkapp and Berlevåg in Finnmark). Of interest, when one recalls the rivalry which Arnøy people claim exists between the centre (Skjervøy) and the districts (such as Arnøya), is that Skjervøy kommune's trawl licence went to the Årviksand-based Alfredson, and not to a vessel based in Skjervøy. Some people suggested (and I tend to agree) that the decision to locate the trawl licence in Årviksand clinched that community's standing as the real, perhaps even archetypal, fishing village on the island. The trawl licence stipulated that although the Alfredson could be sold outside the kommune, the trawl licence could not. The importance of having a trawl licence was explained by one Arnøy fisher as follows, *the meaning of the trawl licence is that you can fish in the periods that conventional boats don't. So, the fish productions [fish plants] could go [work] all year, with cod. The reason that the Alfredson, the boat, still [in 1997] is in this kommune is because of the trawl licence, the paper from the Government, which says that the Alfredson can fish with a trawl. That paper is given to one boat and that boat has to be from Skjervøy kommune. So you can't have the boat out of the kommune. And that's the reason, because they [the brothers] have the boat. The boat can be sold outside the kommune but you don't get the paper giving you permission to fish with a trawl.*

However, even with their new trawl, during the years immediately following, declining fish stocks netted the Alfredson only poor catches and escalating debts. The brothers offered to sell the Alfredson, as its skipper explained: *in 1989 [sic, 1990] the quota system began and then the crisis began. That's why the boat was for sale. And also the [scientific] researchers said that there was much cod in the Barents Sea but there wasn't you know, because of wrong [scientific] counting. Then between 1989 and 1992 they was building up the cod [stocks].*

Around 1988 or 1989 [one person said 1987], the Alfredson went bankrupt for the first time. Some said that an Akkarvik-based Cooperative called the Arnøy Sea Fishing Company bought it. A quixotic Oslo accountant married to an Akkarvik woman spearheaded this Cooperative. The accountant envisioned buying other boats as well, and planned to buy fuel and other necessary fishery supplies in bulk, so as to supply local fishers with cheaper necessities. One man explained the accountant's goal: *his thought was if you bring the industry, the boats, together, you can get more money, the cooperative system. You've got more boats in the company so the boats have somewhere to deliver fish and the producers know that they will get it. It was a [good] thought but it was very bad times in the fisheries.*

At about the same time, the Skjervøy Sea Fishing Company, which owned several large fishing vessels, also went bankrupt. The Akkarvik co-op hoped to buy its vessels but lacked the means to do so. Instead, the powerful Tromsø-based firm of Kræmer purchased the Skjervøy vessels. Moreover, closure of the fishery on 18 April 1989 appears to have put an end to the Akkarvik co-op. The Alfredson was repatriated back to Årviksand prior to the beginning of the closure of the fishery.

The Norwegian Fishery Directorate closed the fishery on 18 April 1989 and reopened it on 1 January 1990. That the Alfredson was fishing from Årviksand on the eve of the fishery closure is clear from the words of a long-term crewman. When I asked him about the mood of people when government closed the fishery, he responded, *people were depressed. When it stopped in 1989, people were [ironically] catching much fish. They didn't believe the government. They were fishing in this area. I was with the Alfredson.* As described below, the people of Årviksand rescued the Alfredson from bankruptcy or from potential buyers, including the Akkarvik co-op. In the skipper's words, *the current mayor of Skjervøy [originally from Årviksand] ... the boat was for sale, the mayor and the local society [local people], they helped, I mean economically, to keep the boat in this place. From the politics of Troms Fylke, they have said that as a fishing community, Årviksand must have a fishing boat. That's why they had to have boats in this community, and in Skjervøy also.* The skipper's reference to the local society's involvement in repatriating the Alfredson refers to efforts by Arnøy people, mainly in Årviksand, to invest in the Alfredson, and thus to help to pay its debt. There would later be a second local 'drive' in 1993 to save the Alfredson.

A local man described this first drive as follows, *the first time it [the Alfredson] was sold, it was [the brothers] that owned the boat, they couldn't pay the money they owed. Then the people of this place started an A/S* ***aksjeselskap*** *(corporation), many people put the money in a pot, in 1989.* In Årviksand, the 1989 drive to save the Alfredsen required a door-to-door campaign to raise money. One person, who dates the incident to a year later, described the campaign as follows, *the people of Årviksand and Skjervøy* ***kommune*** *purchased an* ***aksje*** *(a part or share of the boat) for 25,000 NOK. The bank said we could take the boat back because people in* ***kommune*** *explained that Årviksand must have it. The Alfredson was taken in March 1990 and returned in October 1990.*

However, three years later, in 1993, the Alfredson faced new problems, this time caused by an increased debt from buying new technology, and a poor fishery. One person summarized some of the problems faced by the vessel between 1989 and 1993, as follows: *the first time it was sold, it was Einar and Arne that owned the boat, they couldn't pay the money they owed. Then the people of this place started an A/S* ***aksjeselskap*** *[corporation], many people put the money in a pot, in 1989. The reason we lost the boat in 1993 was the engines. The boat stood still for 3 or 4 months, and then [they] put a new engine in 1992. That was a big expense, the new engine cost about 1 million. In 1993, we worked to buy the boat back again, Skjervøy* ***kommune*** *and Arnøybruket [the Årviksand fish plant, see below] paid money to buy the boat back again.* Another person put it this way, *they [the brothers] bought a new motor and it cost very much. Then, they bought some fish-finding technology and it was costly. They didn't catch enough fish to cover these expenditures. It was 5 or 6 millions they had in loan. Today, I heard, I'm not sure, but I heard, it was 3 million. If they catch enough fish, they could pay off the loan.* Another fisher echoed the fisher just quoted, attributing the difficulties the Alfredson had paying for the renovations to low fish quotas in 1993. He said, *the quota was not high enough in 1993 to catch enough fish to support the boat. There was not enough quota to make money to afford an Alfredson. So when they bought it back, the quota was higher.*

Once again, crewmembers of the Alfredson literally went door-to-door to raise money to save their vessel, but this time, some locals were more reluctant to dig deep into their pockets to help. The passage of time, the fishery crisis, out-migration, and withdrawals from the crew's account of social capital, led to a different outcome. As one man remembered, *in 1993, the bank again took the Alfredson. And then they (the skipper and crewmen) came around to the houses and asked the people to buy* ***aksje*** *(shares), but Daddy said he could not again.* [John: But some did invest?] *Yes. And again the* ***kommune*** *gave money and Arnøybruket [See below] gave money. And people from Skjervøy.* Another person described the 1993 drive in Årviksand as follows: *Einar was on the south side of the river; Arne solicited on the north side. Knut talked to his brother, and he gave money.* [John: those giving became shareowners?] *Yes. My dad, he worked on the Skjervøy* ***kommune*** *[council] to get money from them. They gave 100,000 NOK as did Arnøybruket. It needed 800,000 NOK.*

By this time, the saga of the Alfredson had attracted the attention of the Tromsø media. A 23 October 1993 story in the Tromsø newspaper Nordlys [RL, translation] reported that, *the Alfredson refused to give up. Everyone sacrificed something to save the [community's] life's nerve, the fishing boat. Money collection to save boat. Årviksand and the kommune lost the boat. [The skipper] lost his boat, his work place. Now he and seven other crewmen live on 250 NOK per day in unemployment payments. The fishing community of Årviksand keeps up hope. People go door-to-door collecting money. The Alfredson was Årviksand's last big vessel and a secure deliverer of fish. When it was sold, a part of the life nerve of the community was severed. If there were free places on other boats, [the skipper] and his crew would have to take them .... The harbour at Årviksand is empty of larger boats, only small* ***sjarks****. [The skipper] is very worried about the future of*

Årviksand and wants to buy the 80-foot [sic] Alfredson back after the forced auction. The auction has not occurred.

Over the next few weeks several buyers expressed interest in the Alfredson but in mid-November (1993), Skjervøy mayor Kurt Leif Strøm (originally from Årviksand) said that the Alfredson, and specifically, its trawl licence, should be given back to Arnøya and threatened to call in Norwegian fisheries minister Jan Henry T. Olsen[8] if the boat was sold elsewhere. Five days later, on 18 November 1993, the brother of the Alfredson's skipper, who had won one million NOK in the lottery two years before, decided to contribute his lottery earnings to help pay the Alfredson's debt. He was quoted in the Tromsø newpaper Nordlys as saying; *I've lived by fishing and will continue to!*[9]

One of the fishers I interviewed summarized the Alfredson saga, and its relation to the trawl licence as follows: *it is the trawl licence which brings that boat back to Årviksand each time.*

Fish Plant Turnovers

As we saw in chapter 3, during the years immediately preceding the crisis of 1989 there were at least three fish plants on Arnøya. Leinebø and Larsen operated the plants at Akkarvik and Årviksand, and Rolf Pedersen operated Skaretfish at Lauksletta. This situation would change significantly during and immediately after the crisis beginning in 1989 and as in the case of the vessel Alfredson, the political importance of employment figured significantly in the rise and fall of fish plants.[10] I begin with Årviksand.

Årviksand

As we saw in the last chapter, Leinebø and Larsen maintained its fish filleting plant at Årviksand, surviving the mandatory closure of the fishery in 1989 and hobbling along until October of 1990. One person who worked in the plant at that time explained that, *the period of turbulence started in late 1989, there was a period of time when the plant closed because the Norges Råfiskerlag [Norwegian Raw Fish Association] took the buyers licence from Leinebø and Larsen, because the company lacked sufficient money to ensure that the fishermen got paid. The factory was closed for a period of time.*

---

[8] Norwegians call Olsen 'no fish' Olsen because of his reluctance to grant fish quotas to EU countries.

[9] Information in this paragraph comes from two *Nordlys* articles published on 13 and 18 November 1993. I thank RL for showing me the articles and translating them.

[10] Respondents in my adult questionnaire ranked the fishery (38 percent) and fish plants (33 percent) as the top two economic developments to affect Arnøya between 1986-95; the political clout of such opinions is obvious.

In early 1991, the Tromsø firm of Norfra leased both the Skaret plant in Lauksletta and the Årviksand fish plant from owners Leinebø and Larsen. Norfra (meaning 'from the north') was a small company, formed by a fish exporter in 1984. Initially, Norfra concentrated on exporting fish (mainly to France) and on salmon aquaculture. For three months (October to December 1990) preceding Norfra's lease of the Lauksletta and Årviksand plants, there had been no fish plant operating on Arnøya. The Norges Bank stopped credit to Leinebø and Larsen and Norges Råfisklag terminated their buying licence, because the old company lacked money to pay fishers for fish.[11]

Under the new lease, Norfra employed Leinebø and Larsen's Årviksand plant manager and most of its previous employees. Prior to Norfra's leasing of the Lauksletta and Årviksand plants, and indeed during it, considerable uncertainty surrounded the future of both plants. Årviksand people worried that if filleting jobs were lost, more families would leave Arnøya. Yet Norfra continued to fillet fish at Årviksand, and renamed the old Leinebø and Larsen plant 'Arnøybruket', the Arnøy fish factory. Meanwhile, over at Lauksletta, Norfra ran Skaretfisk as a plant concentrating on more traditional salting or hanging of fish.

During the period of Norfra's lease, a larger, more established fish processing company, Kræmer of Tromsø, expressed an interest in purchasing Årviksand's fish filleting plant. Local support for Norfra and Kræmer was divided during this period of indecision and division. One Årviksand man described this time as "like a nightmare in the daytime," as follows: *there was much uncertainty. Several companies wanted to buy the [Årviksand] factory. For example, Kræmer in Tromsø wanted to buy the factory. There was a meeting of the workers and people living in Årviksand. In a way it split up the citizens of Årviksand. Some supported Kræmer, some Norfra. At that time I didn't like being in Årviksand. This was late in 1990 or early in 1991. But it was before the take over. Kræmer is an old company, well known. I knew people who worked on their trawlers and they had nothing to complain about with the company. Norfra was newer. All we knew is that they sold salmon to the French. A lot of people were sceptical. So, I had my house and my family to pay for. And at that time (1991), plenty of people were saying that they don't believe in the Norfra people, don't believe they have the money and maybe they just want to buy the factory and then sell the equipment. There was a lot of turbulence around the whole thing. I had my job taking care of the frozen fish we had. But then the uncertainty started. Before that, everyone thought they had safe jobs, just everyone else had to leave their jobs. Suddenly it came, like a nightmare in the daytime.*

Another Årviksand man recalled the village mood of that time, *they say [i.e., said then] Årviksand pays only for the old [pensioned] people, the young are going away... It was a dark mood, a difficult time.* By this time, Kræmer had become

[11] Henriksen (1990) provides some details from this period, as presented in this paragraph.

majority owners of Skjervøyfisk, a filleting plant at Skjervøy, and this fact helps understand why the debate on Arnøya over whether Norfra or Kræmer would be better for Årviksand revolved around each company's plans for just *how* it would process fish.[12] One man recalled that *Norfra exports, after 1989 they had lots of money. [There was] much local discussion of the improvement. Leinebø and Larsen, who preceded [Arnøybruket] had poor economy. People were watching closely. Norfra had money in salmon farms, exports, and had [a fish plant at] Torsvåg. People were initially sceptical. There was another company from Tromsø [i.e., Kræmer], very old and very big. They were interested in the factory as well. The people chose between the two companies. We heard they [Kræmer] would only land fish at Årvik and transport it to Skjervøy. Norfra had another policy. They wanted to make fillet, and so on and that's why we chose Norfra.*

By September 1991, the concerns of Årviksand people over the future of their plant attracted national attention. The Norwegian secretary of state for fisheries, Per Alf Andersen, publically denied that there was a hidden agenda to abandon all Arnøy fish plants or that the government favoured the Kræmer bid to buy the Årviksand plant.[13]

Arnøy people were concerned about a Kræmer takeover of the Årviksand plant for two reasons. First, since the late 1960s, the Norwegian government often paid fish plant owners to stop production at local plants, so as to decrease the fishing effort. Second, landing all fish from the area at Skjervøy and filleting it there was not only feasible, but would allow larger vessels, including ships from Russia to land fish (See Chapter 5).

Several public meetings were held in which the Norfra and Kræmer bids were debated. One of these occurred in January 1990 and was characterized by one fisher as a *referendum, in which everyone voted and included presentations by politicians from Skjervøy and Tromsø*. The final showdown came in September of 1991. Arnøy people were afraid that if Kræmer took over the Årviksand plant, all filleting work would move to Kræmer's Skjervøyfisk plant in Skjervøy, reducing Årviksand's fish plant work force from 30 to 6 or 7. Two hundred people signed a petition against Kræmer taking over the Årviksand plant.[14] In the end, Norfra purchased the Årviksand plant on 4 January 1992.

---

[12] A two-person delegation from Årviksand traveled to Tromsø to interview Norfra and Kræmer management. One reason why locals preferred Norfra was that it, unlike Kræmer, owned no nearby plant where equipment from Årviksand might be moved.

[13] The Norwegian government's alleged role in Årviksand's fish plant decision is discussed in Henriksen (1991a)(RL translation).

[14] Locals recalled details of the September petition and Henriksen (1991b) (RL, translation) also describes the event.

From the beginning, Norfra's plan for Arnøybruket emphasized the exportation of chilled fresh fish, along with some filleting.[15] Norfra's export strategy built on the company's export tradition and was enhanced by a 1990 change in legislation which sidestepped middlemen, allowing fish processors to directly export their own products.[16] Norfra's General Manager explained the company's export strategy as follows: *you have to start with the fish, once they come aboard. The top quality fish can be sold for highest price. [We] bring top quality fish down to Europe in 50 hours.*

In order to make the Årviksand plant more efficient, in the autumn of 1992, Arnøybruket sought help from Skjervøy *kommune* to extend the dock in front of the old Leinebø and Larsen plant, and the following year, Norfra used federal funds to modernize several sections of the old Leinebø and Larsen plant, including the salt fish room, offices, and staff room. In August 1993 the federal Fund for Trade [and business] in the Districts (*Statens Distrikts og Næringsfond* or SND) promised Arnøybruket 2.2 million NOK to develop new production buildings, which the Tromsø newspaper *Nordlys* described as *part of the great building of Arnøybruket, a daughter company of Norfra.* These and other changes to Arnøybruket began in January 1994. The cost of these initial changes was considerable (approximately 6.5 million NOK) and shared between Troms Fylke, SND, Skjervøy kommune, bank loans, and Norfra. The Arnøyhamn-based construction firm of Sørensen and Sønn undertook much of the construction, multiplying the local economic benefits. Arnøybruket (not its parent company, Norfra) also acquired minority ownership in a number of local fishing vessels, including forty per cent ownership of the Alfredson.

Akkarvik

Leinebø and Larsen continued landing fish at Akkarvik and producing dried fish (*tørrfisk*), at least until closure of the fishery in April 1989. In hindsight, one might question the ironic fact that in 1988, the Fiskeridepartementet spent 1.3 million NOK adding a new north wing to the Akkarvik Plant. Two years later, in 1990, Leinebø and Larsen accepted 1.1 million NOK from the Fiskeridepartementet to cease producing fish at its Akkarvik plant. However, during the time discussed here, Leinebø and Larsen still owned the plant and the 1990 buy-out did not (in theory) prevent fish from being landed there (subject to necessary wharf repairs), for processing elsewhere. Yet the buy-out ended the company's right to produce fish there, with a loss of employment and economic infrastructure. An Akkarvik fisher explained the buy-out this way: *when Leinebø and Larsen owned it, they could land and produce fish there. However, they got paid a million to stop*

[15] Norfra's highest quality prawns and other fish products are marketed internationally (although mainly in Europe) under the name of Grand Nord.

[16] The 1990 change in export legislation is mentioned in Apostle et al (1998: 285).

producing and they'd get one half a million [more] if they didn't take [prohibit fish to be landed] the fish also but the Norges Råfisklag managed to stop that. We [local people] got a message to the national Råfisklag so that they can land fish here but now, it's not good enough [wharf facilities]. Toward the end of his remarks, this fisher is alluding to the fact that the wharf facilities at Akkarvik deteriorated between 1990 and 1995 to the extent that fish now are no longer permitted to be landed there. Although the Akkarvik Fiskarlag attempted to raise money to improve the wharf so as to permit landings, their efforts have thus far been unsuccessful. And with the loss of Akkarvik's salmon farm in 1997, Akkarvik, Arnøy's best natural harbour, lies idle (See Epilogue).

Lauksletta

The fishery crisis claimed the Skaretfisk plant at Lauksletta in the summer of 1989. Years before and until the fishery crisis of the late 1960s, there had been two small, locally owned fish plants in Lauksletta when one of them, the Nygård plant, accepted a financial package from government to stop production. The other plant, Skaretfisk, owned and managed by Rolf Pedersen since the mid-1960s, survived the crisis of the late 1960s but finally succumbed to bankruptcy during Norway's short fishery moratorium of 1989. Skaretfisk had had a small filleting operation but relied primarily on the traditional hanging and salting of fish. Rolf Pedersen had a difficult time seeing the plant he had run for 25 years change hands. He attempted to persuade a Karlsøy fish plant owner to make an offer, so as to keep his hand in the plant. Also, while Skaretfisk was in the hands of the bankruptcy board (*Konkursbo*), some Lauksletta locals attempted unsuccessfully to purchase the plant. In the end, the only offer to buy Skaretfisk came from Tromsø-based Leinebø and Larsen, who enjoyed backing from Skjervøy *kommune* and the banks. However, Leinebø and Larsen's main interest in Skaretfisk appears to have been to qualify for government monies to cease production, which occurred in 1990. Although, as noted above, Norfra briefly leased Skaret in 1991, the company focussed mainly on its Årviksand plant. Consequently, over the next two years, there was no place for landing or producing fish in Lauksletta. During this difficult time, Lauksletta fishers transported their own fish, sometimes in the trunks of their cars, 18 km to Arnøyhamn, which at the time, served as a collection depot for fish that was then trucked to Skjervøyfisk in Skjervøy.

The Lauksletta Bygdelag discussed the fact that there was no place to land fish in Lauksletta during their April 1992 meeting, and this, along with other local lobbying, may have convinced *kommune* politicians of the need for a plant at Lauksletta. Meanwhile, Rolf Pedersen, the former owner of Skaretfisk, who one local person described as having *political connections in Skjervøy, Tromsø, and Oslo,* continued to look for a financial backer to help him regain some control of his old plant.

His patience finally paid off. In November of 1992, Rolf Pedersen, with financial backing from the Karlsøy fish plant owner he had approached earlier, reopened Skaretfisk. A 10 November 1992 article in Nordlys (RL, translation) pictures Rolf Pedersen, noting that his new Skaretfisk has the necessary buying licence from Norges Råfisklag and guarantees from the bank of sufficient capital to pay fishers. The renewed plant opened shortly thereafter, and, as reported in Chapter 5, now buys fish both from local fishers and from Russian trawlers. The new Skaretfisk, now managed by Rolf Pedersen's son, plays an important part in the Arnøy fishery.

With origins both in the sea and on land, Norway's short-lived fishery crisis was an important turning point in the recent history of Arnøya, with significant consequences for each of the island's villages. The loss of economic capital at Akkarvik, chiefly its fish plant, meant that remaining Akkarvik and Langfjord fishers had to offload their fish at Årviksand or at Arnøyhamn for trucking to Skjervøy. More importantly, the Akkarvik plant workers lost their place of employment. Some people left the community, others either moved to Årviksand, or commuted there daily to work. One community's loss was another's gain. Akkarvik's loss of its fish plant led to out-migration, especially of young families. Those who remain are increasingly older, and overall, out-migration reduces the pool of people available to participate in local organizations, such as the *bygdelag*. Those remaining valiantly protested the loss of their plant, and the opportunity to land fish, but their declining numbers underlined their struggle.

Årviksand emerged from the crisis with the loss of four of its five large fishing vessels. By 1995, the ability of its small, local fleet to regularly supply fish to Arnøybruket was uncertain. As man-made behemoths that require regular feeding, fish plants require a steady diet of fish to operate and this need is local and economic, as well as regional and political. With its relatively unproductive small boat fleet and the loss of its larger vessels, and the community's renovated fish plant, this need for fish intensified pressure on the Alfredson to produce. For some people, the Alfredson and its crew could never fish long nor successfully enough and by the time of my fieldwork the boat may have become the scapegoat for the local fleet's lack of productivity. Although such sentiments were neither universal nor fair, there was also a degree of resentment about Alfredson's trawl licence, even though anyone expressing such an opinion would, if questioned, readily acknowledge the community's dependence on the vessel. Local ambivalence toward the Alfredson, both criticism and dependence, was manifest in the local response to the second door-to-door drive to prevent the vessel from being sold from the community.

Moreover, as discussed in the next chapter, Årviksand's shallow, artificial harbour prevents larger fishing vessels, such as Russian fishing vessels, from landing fish there. But viewed differently, Årviksand's strength during the crisis was evident in its support for the Alfredson, at least during the first door-to-door drive to sell shares in the vessel. Although the community saved the vessel during its first bankruptcy, whether people were motivated by trust and good will, the gist of social capital, or by the realization that they and their neighbours needed the vessel, is more difficult to determine. Assuming that some degree of trust and good will permeated the first drive, such sentiments diminished by the second, requiring the skipper's brother to use his lottery earnings to save the vessel. I also contend that the debate over whether Kræmer or Norfra would operate Årviksand's fish plant was, from the distance I write about it, a healthy dispute, and one less likely to occur in Canadian fishing communities. Although heated while it lasted, any scars from the dispute had healed by the time of my research, and my sense is that worker dissatisfaction with Norfra, one of many themes reported in the next chapter, is neither widespread nor a vestige of the ownership debate.

Over at Lauksletta, the revival of Rolf Pedersen's Skaretfisk was very different than the emergence of Arnøybruket at Årviksand, chiefly because Pedersen himself was the dogged proponent of his firm's return, rather than the community itself. Indeed, the community-based initiative to gain control of the bankrupt firm failed. By the time of my return to Arnøya in 1997, Skaretfisk was operating successfully, even though most of its workers still commuted from Skjervøy, and were thus not from Lauksletta. While located in Lauksletta, Skaretfisk was only partially a part of the village's economic infrastructure, as discussed more fully in the next chapter.

CHAPTER 5

A VOYAGE TO FINNMARK: THE CONTEMPORARY FISHERY

[Notes from my diary introduce this chapter on the contemporary fishery, and begin aboard the Alfredson during a five-day fishing voyage to Finnmark].

.........Berlevåg, Finnmark, 5 May 1995. Today I recorded a daily round aboard the Alfredson. The Alfredson and its seven-man crew came into Berlevåg last night to sleep dockside, having fished around the clock for two days, enjoying good catches. The exhausted crew turned in around 8:00 PM.

Today begins at 3:20 AM. The engines have been turned on. I lie in my tomb-like bunk, located at the bow of the ship, on the lower deck, relatively inaccessible from the main living quarters and work area, located one deck above and near the stern of the ship. We depart Berlevåg about 4:00 AM, steaming though rough seas for about two hours to the place where fishing will resume.

Turbulence rules out sleep during these two rocky hours. Every time the bow lifts and crashes again downward, my body lifts a few centimeters off my hard, narrow mattress, and throws me down again, like some outclassed wrestler. I am at the stage where if I were a person who got seasick, it would be now. I am not. Times like these make me wonder why I asked to make this trip. I am entertained by an orchestra of sounds: sudden clangs, anguishing groans, and behind all this, the constant drone of the engine. I appreciate anew the value of fish and those who catch them.

My tomb is accessible enough in calm waters, but this morning's seas further separate me from the crew. To reach them and the ship's bathroom, I must first get out of my bunk, easy enough in calmer seas, or for a hardy Olympian, but a real challenge for **me** *this morning. I must then climb over various plastic fish storage cases, over and around various paraphernalia in the fish cutting room, over fish entrails (some remain even though the room is washed down after each catch) and eventually to the steep, narrow stairs leading to the upper deck.*

The first trawl is set around 6:00 AM. From then on, the crew works non-stop. They set the trawl another three times today, unload the catch each time, then descend to the fish cutting room to cut, and pack fish in plastic cases surrounded by ice. They stop for around 20 minutes for supper at 7:15 PM, before eviscerating the final catch of a day, which began before 4:00 AM and ended after 9:00 PM.

The vessel has three decks. The top deck is primarily open workspace; most of it is in front of the cabin from which the skipper or another crewman pilots the vessel. The gigantic cylinder on which the bottom trawl is wound and stored fills

Arnøybruket. 1995. (Photo by the author)

the top deck behind the cabin. The floor below the top deck can be divided into two main portions, one forward portion is mainly a fish cutting room where fish are processed. Cod and haddock are beheaded and eviscerated, then packed in ice. Fish heads and entrails are discarded into the ocean through a stainless steel chute where they attract an assortment of seabirds. The rear portion of the second deck is primarily crew living space. There is a small kitchen, with double sink, stove and storage cabinets. Opposite is the crew's common room containing a large table surrounded on three sides by padded bench seats. Here, all eat, smoke cigarettes, and drink coffee. The room also contains a colour television, radio, and storage shelves. On the basement floor, we have first the crews' sleeping quarters in the stern. Another similar ladder from the second story hall leads down to the engine room, located forward of the crews' sleeping quarters. The next section forward is the hole where fish is stored in plastic cases, a little over a meter square (iced top and bottom). Finally, on the bottom floor at the bow is another bedchamber, with space for 4, two bunk beds, including my own.

My trip aboard the Alfredson started on 1 May. Knowing the Alfredson was in Årviksand, I decided to ask the skipper if I could accompany them on their next trip. I arrived at his door just as he was leaving, likely to visit one of his brothers. My arrival changed his plans. He accompanied me inside, took off his coat, and ushered me into the living room where his wife joined us. She soon served ***lefse*** *(thin, folded pancakes filled with goat's cheese and sugar) and coffee. We communicated through a medium of my paltry Norsk and their deliberate Norsk, interspersed by occasional words in English. I finally asked if I could accompany* *the Alfredson tomorrow. 'Yes' the skipper immediately answered, adding that they would leave tomorrow at 6:00 AM, spend part of the day in Skjervøy repairing the trawl, and then head north to the Finnmark coast. Their adult son appeared half way through this conversation and his English clarified some points. The skipper asked if I had coveralls and his son volunteered the loan of his. I left around 11:00 PM and told my surprised wife. We quickly packed, set the alarm for 5:20, and turned in.*

Early next morning I joined the seven man crew, comprised of Einar, 52 year old skipper, his 54 year old brother Arne, engineer Trond, 54, cook John, 51, and his 23 year old son, Svein, the second engineer, Knut, 42, and Tormod, aged 27. Svein and Tormod speak some English, while the skipper and the rest understand some English.

En route to Skjervøy (about 2 hours, at about 10 knots) I speak with Svein about the opening of the Norwegian minke ***(vågehval)*** *whale hunt in the Barents Sea. As we steam into Skjervøy, Svein points to the Tromsø-based oil carrier, the Norsel, a vessel which carries oil to Icelandic (and Norwegian) fishing vessels at sea, so they do not have to return to land. Svein explains that Norwegian fishers don't like the Norsel because it sells oil to Icelandic fishers, whose fishing practices they consider predaceous. We spent most of the day at Skjervøy, first at Refa Fiskeredskap, repairing the trawl; next to purchase 2500 NOK worth of groceries including meat, potatoes, and other root crops, bread, coffee, and so on, for the trip. Each crewman also shops for personal supplies (candy, tobacco, potato chips, bottled 'cider', soft drinks, and others) while we wait for half an hour to take on 10,000 liters of fuel. Finally, we go to an industrial shipyard to have some changes made to the trawl cable and new bottoms welded onto the trawl doors. To my inexperienced eye, the changes made to the trawl appeared a mix of hocus-pocus and science. In this case, the trawl was fishing off-center and, therefore, not following directly behind the boat. After some discussion between the skipper and shipyard workers, one side is shortened. Throughout all this, the skipper either completed or coordinated much of the work. He appeared impatient throughout, cognizant of the fact that employees at the shipyard needed to remain after closing time (3:30 PM) in order to replace or reinstall the steel cables towing the net and to repair the trawl's massive steel doors.*

We finally left Skjervøy at 7:00 PM, after an eleven-hour stop, and headed north across the mouth of Kvævangen Fjord, and then followed a route protected by offshore islands. I chatted with Tormod about the fishery. He said the money was good, prompting me to ask why more young people were not in the fishery. He answered that until the 1989-90 crisis, more young people were in the fishery but dropped out at that time. Tormod believes young people are now returning to the fishery.

The journey from Arnøya north to the spring fishing grounds off Finnmark takes around 20 hours, at about 10 knots. The time en route is consumed by discussions of fish, football, the weather, and other topics. Crewmen consume endless mugs of coffee and most roll one cigarette after another (only 2 of the 7 member crew do not smoke). Relations between crew members seem remarkably good given their close working relationship during the nearly eleven month fishing year. The skipper appears more removed, perhaps because of his responsibilities; although all crewmen spend time at the helm, his time exceeds all others.

In 1995 the Alfredson had a quota of 280 tonnes of cod (*torsk*), head off and gutted; 80 tonnes for haddock (*hyse*); 1,100 tonnes for saithe or coalfish (*sei*); and 1.3 million kilos, for herring (*sild*). Individual vessel quotas (or IQs) were introduced to the Norwegian fishery in 1990 in response to the fishery crisis discussed in Chapter 4.[1] Quotas were initially set according to vessel size and on the basis of catches during the three previous years. Quotas change annually, reflecting both the size of the vessel, catches, and scientific assessment of stocks. On the other hand, quotas can decrease. Increases allocated to the Alfredson since 1990 reflect both the vessel's actual catches and improved stock assessments. For example, in 1990, the Alfredson's quotas were 50 tonne for cod; 200 tonn for herring; and 900 tonn for saithe, far less than the vessel's quota five years later. In 1995 there were *no* quotas for prawns (*reke*), red fish (*uer*), and catfish (*steinbit* or *gråsteinbit*).

Individual vessel quotas are a new form of economic capital. Vessels with (relatively) large quotas, such as the Alfredson, require larger crews, and consequently, employ more fish plant workers ashore than do smaller vessels with lower quotas. Even so, my sense is that the introduction of the IQ system has increased mistrust and envy between fishers, and has therefore decreased social capital.[2] Many of the fishers I spoke with believe that larger vessels were favored

[1] Unlike Iceland's system of Individual Transferable Quotas (ITQs), which permits quotas to be bought and sold (See Pálsson 1993), the Norwegian IQs, like most vessel quotas in Canada, cannot be sold.

[2] In my *figurative* use of the term *social capital,* the term refers to social assets, analogous to economic capital, which some individuals possess, and are based in the trust they experience with others. In my view, the *social capital* concept becomes more problematic when extended to groups or communities, because, among other weaknesses, measuring degrees of social capital becomes an inescapably subjective exercise. The classical origins of the term may be traced to Marx and Durkheim; especially the latter's concentration on social integration. More recently, social capital is for Bourdieu (1985) non-monetary, instrumental, and socially constructed; actors deliberately create social capital to gain access to other capitals -- human, cultural, and institutional. On the other hand, Coleman's (1988) social capital builds on reciprocity and social exchange theory, noting how an actor's access to social capital incurs a debt to those in their social network offering it. Elements of Bourdieu and Coleman are apparent in Lin's analysis (2001). Perhaps because it discusses the 'eclipse of community,' Robert Putnam's work on social capital has attracted considerable attention among students of small and/or rural communities. Putnam's account of social capital begins with his 1993 analysis of the contrasting roots and expressions of 'civic society' in north-central and southern Italy. Putnam attributes the rise of a more effective democracy in the north to a higher incidence of social capital stretching back to medieval communal life. Social capital is based on trust and exemplified by volunteerism. Critics of

over small ones when the system began, fueling inequities between fishers. Overall then, while most Arnøy people believe that the introduction of individual vessel quotas was a good idea,[3] and necessary to conserve fish stocks, many believe that small boats should have unlimited or larger quotas than large vessels, a position of considerable merit.

Fishing

As we learned in chapter 4, the Alfredson's trawl license provides the vessel with a coveted advantage over most other island fishing vessels. And, while the Alfredson employs other fishing technologies when advantageous, the trawl is the cornerstone of the vessel's fishery. The trawl used is called a *bunn trål, bunn* meaning bottom. The trawl is a net bag resembling a large stocking towed behind the boat. Viewed more critically, it is an proficient fish-killing apparatus, knit from green nylon netting, weighted down by iron doors, which measure about 1.5 m by 1 m, that serve to keep the trawl mouth open while the trawl is pulled along the bottom. One kilometer-long steel cables tow the net. Weighted rubber bulbs on the bottom of the trawl aperture bounce along the ocean floor. An assortment of chains and weights, and buoyant plastic floats, which keep the top of the door up, complete the trawl. Such trawls occasionally catch debris, ranging from boulders to autos (a trawler brought up a car 3 May 1995 -- because, people claim, Russians dump cars into the sea, after pirating parts from them).

When autos, boulders or other sharp objects cut netting, crews use fish knitting needles to mend holes. When the trawl is being hauled in, one hears a chorus of clangorous sounds, including, at the beginning, the whirring of the heavy iron doors being raised to the surface. This takes about 10 minutes. So as to reduce the resistance of raising and securing the doors, the skipper reduces the boat's speed to less than one kilometer per hour. The unmistakable sounds of the trawl being brought in are soon followed by the appearance of the crew, converging from various locations to assess the size of the catch and prepare for work.

Once raised from the water, the trawl's doors are unhooked from the wire leading to the net, to which they had been attached and securely fastened to the ship. Once the doors are up, hoisted by the same cable which will soon raise the net, and stored on each side of the boat's stern, they are fastened with chains and secured with rope. The skipper then increases the speed to near trawling speed (e.g.,

---

Putnam (1993), such as Portes (1998) note (I believe correctly) the theory's logical circularity, insofar as social capital is both the cause and effect of what is being explained. More recently, Putnam (2000) attributes the decline of 'community' in 20th century America to reduced civic participation. However, increased expressions of social capital provides the 'magical' instrument for the revival of community, although just where the trust will be found that Putnam argues is necessary for rekindled collective action is less clear.

[3] Sixty percent of people participating in my adult questionnaire either agreed (46 %) or strongly agreed (14 %) that the introduction of the IQ system had been a good idea.

Trawl net beside Alfredson. 1995.
(Photo by the author)

about 2.5 kph), in preparation for the winding in of the net. All eyes now focus about one hundred meters or so behind the boat, where the net bag appears on the surface. The net has surfaced because of the raising of the doors, which have hitherto helped submerge it, and because of the buoyancy of the fish and plastic floats. The appearance of the net bag is colourful and exciting. If redfish (*uer*) are part of the catch, their colour contrasts with the green nylon of the net, producing a yuletide-like display of greens and reds. Crewmen amicably debate the size of the catch. Next, the skipper begins winding in the net and floats onto the large iron spool, which stores them. This winding produces a din of jingling sounds, as the chains are wound around the spool. The ever-decreasing sweep of the trawl net is now on the surface of the water playing out behind the boat.

Days later, over coffee and in the comfort of his living room, the skipper and his son explain further details to what I have just described. *The* ***bunn trål*** *goes right on the ground. The metal doors keep it down. The trawl is about one kilometer behind the boat. Once the doors are [pulled] in, the whole net is closer to the boat, floating up. Air in the fish makes the trawl rise. Fish bladders explode when they come up from bottom. The organ, which explodes, adjusts to different depths. It's called the* ***svømmeblære*** *(air bladder), in Norsk. Take a fish on the bottom; if you take the fish fast up, the organ explodes, that's why it floats. When the trawl is up and close to the boat, I take the boat in a clock-wise direction, to keep the trawl* *clear of the propellers. The trawl is then alongside the boat. Activity moves to the front of boat. The net bag is at the end of the trawl. A rope goes round the net bag. A slipknot goes one time around the net bag. The coast guard has determined that the slipknot rope be 248 cm. The mesh in net bag is 135 millimeter, stretched. The end of net bag has cod linea knots holding the bag secure. Fish cannot drop out.*

The seven-man crew of the Alfredson works long hours. The skipper is in charge of all phases of fishing. The skipper must record details of each trawling, showing when he began, ended, amount caught, and so on. All this information is kept on record for the office of fisheries (headquartered in Bergen). The skipper is also responsible for the operation of the vessel and decides when and where the trawl is set. When I asked about how he knew when and where to set the trawl, he answered uneasily, *I've learned it. [Then, laughing] ...I have a good background in [knowledge of] the area and exchange information from other fishers. And the maps also indicate good and bad possibilities. And I know that the trawl gear is adjusted properly. If it is adjusted wrong you can forget the fish.* [4]

Just prior to my voyage on the Alfredson, the vessel was equipped with computerized equipment allowing the skipper to record catch locations and other details, such as where they have encountered troublesome objects (rocks, car wrecks, or other debris) on the ocean bottom. The skipper directs the crew's activities. However, given that most of the vessel's crew have served for many years with the skipper, his orders are more collegial than imperious, perhaps approaching the 'ideal' skipper-crew relationship.[5] All of the crew fulfilled multiple tasks. The two engineers alternated for six-hour shifts, with responsibility for checking on the engine. The other members of the crew are all on duty at the helm, for three hour shifts, and then off for six hours, on a rotating basis. The skipper's brother (and vessel co-owner) ran the fore pulley, which lifts the filled trawl net bag onto the vessel, while John, the cook, manned the aft pulley. Knut, assisted by Tormod, put fish in cases, in the hole of the vessel. All the crew, including the skipper worked cleaning fish after each catch.

They worked in two crews, each man with specialized tasks: heading, cutting open and gutting fish. Fish heads and guts leave the vessel through an inclined chute on the side of the vessel. Red fish are packed in cases whole; most other species are headed and gutted. In addition to his other duties, John cooks and washes dishes. Ordinarily, cooking begins in mid to late morning, although depending on the level of fishing activity; the meal may not be served until several hours later. During my voyage, for example, cooked meals were served as early as

---

[4]Perhaps following Barth's (1966) discussion of the role of the purse-seine skipper, there has been some discussion about whether fishing skippers (the so-called 'skipper effect') *or* the ever more lethal technologies they employ, determine catch sizes (cf. e.g. Andersen [1972], Pálsson [1991], Durrenberger and Pålsson [1996], and others).

[5] See Andersen (1972: 132).

Mealtime aboard the Alfredson. 1995. (Photo by author)

2:00 PM and as late as 7:15 PM. Prior to each meal, the cook unceremoniously sets the table, tossing cutlery appropriate to the meal beside each bowl or plate. Most cooked meals consist of a gigantic pot of un-peeled potatoes, served with either fish or meat. Men jab their forks into the large bowl of small, boiled Norwegian-grown potatoes. Each man is given a piece of old newspaper on which he places his potato peels, meat or fish bones, or similar scraps. All hold each potato with a fork in his left hand, and uses the knife in his right hand to peel it. Like so much else in Norwegian society there was a remarkable harmony in this process, as if all were peeling to the same drumbeat. Fish is boiled or fried in margarine and served on large platters, often accompanied by gravy of fat and the cooking juices of fish. Some of the cod or red fish served is salted for one to three days, depending on species, then washed and fried. We also ate meat soup, with vegetables -- potatoes, turnips, carrots, and cabbage. Most meals are accompanied by *flat brød* (crisp bread) and *saft* (fruit juice). After consuming his food, each crewmen piles the bones and potato skins which he has accumulated in his piece of newspaper, folds it, and places it on his bowl or plate, along with silverware. Most crewmen remove their bowls or plates to the galley for washing; John retrieves any dishes left on the table.

Fishers described two ways of dividing the revenues from fishing. One fisher who formerly fished on the Alfredson claimed that fishing revenues are divided as follows. In his words, o*n the Alfredson, 54 % is for the boat, 46 % for the crew. All the crew, including Einar, draw salaries from the 46%. Axel Jørgenssen from Skjervøy has the paper [account] for the boat. He took all the money and deals it out. The boat pays for the oil, nets, etc. The crew is only salaries. The boat pays for* *everything else, including loans on the boat. Axel Jørgenssen oversees all this.* One crewmember's description of remuneration emphasized that the crew are paid a percentage of the catch, encouraging hard work. Consequently, as the voyage advanced, the crew's interest increased in counting the tally of fish cases, stored in the ship's hole. The responsibility of keeping this tally belonged to Knut, who posted the current count of cases in the galley and also provided a copy to the skipper. The accumulating tally increasingly became *the* topic of conversation, with Knut staunchly defending his count, even when everyone else thought that the catch had been larger.

As we learned earlier, the spring fishery in Finnmark constitutes an historic and important part of the fishery year of vessels like the Alfredson. In late May, 1995, the Alfredson's skipper described the fishery year as follows: *[We fish in Finnmark until the] beginning of June, until around the 12th of June, still for* ***Torsk*** *with the* ***Bunn trål*** *[Then we have the July holiday]. In August we begin with* ***sei*** *and* ***sild****, in fjords from Arnøya to Vardø. The* ***ringnot*** *and* ***snurpenot*** *are used for* ***sei****. [In late fall] most years, depending on how much we have left on the quota, we might resume cod fishing. But in January and February 1995, with a new [****sild****] quota year we resumed fishing* ***sild****. The* ***sild*** *were big and fat and easy to catch; the* ***sild*** *were standing [stationary] in the Vestfjord.* [John: so the sild is easier or more profitable than the torsk?]. *Easier, because they [the* ***sild****] are waiting [i.e., remaining in the Lofoten Islands] until March or April, when the cod is coming from the Barents Sea. And then the cod is coming to the Finnmark coast, to the coast, and into Vestfjord to spawn. In the last three years there is much* ***sild*** *in northern Norway. This year we saw the few cod come to Vestfjord but encountered many* ***sild****, perhaps the cod are frightened by the* ***sild****. They are either frightened or eating the* ***sild****.* The seasonal pattern of the Alfredson's fishery resembles the rest of Arnøy's large and small vessels.

Other Large Vessels

The Alfredson was Årviksand's lone large vessel in 1995 and 1997 but not the only large fishing vessel on Arnøya. During 1995, the 22.7 meter long Arnøytind, built in 1980, operated out of Arnøyhamn, supplying fish to both the Arnøybruket and Skarbruket plants. The Arnøytind had been skippered by a man and subsequently jointly skippered by he and his son, until the son purchased the vessel in March 1994; the father continued fishing on the boat until April 1995. The next year, the father (then going through a divorce), fished elsewhere and in January 1997, the father purchased the 19-meter Skartind, built in 1974. By 1997 then, the Arnøytind and Skartind supplied fish to Skarbruket (and to a lesser extent, Arnøybruket).

In 1997, the son, skipper of the Arnøytind, described his and his father's fisheries, as follows: *We have [a crew of] eight people [on the Arnøytind]. My father bought a new boat. The young people are on my boat and older people on his, the Skartind. He bought it in January 1997. It is an old boat that worked with*

Sjark fisher unloads catch, Arnøyhamn. 1997. (Photo by Author)

Skarbruket. It was build in 1974... 19 meters. The boat we have [the Arnøytind] is too small for two skippers. You don't get enough money if you fish a half of year each. So, he was with other boats for periods, and changing for skipper on Arnøytind. So, he bought a new boat in January 1997. He was in the North Sea with another boat, in 1996. He fished with our boat until April 1995. I was the skipper since 1995. By 1997, the Arnøytind, with a young crew of eight, fished year round, mainly using seine and ***garn*** *(net).*

The Threatened Small Boat Fishery

As we learned in Chapter 4, the Alfredson, with its trawl licence, was the only large vessel in Årviksand to survive the crisis. On the other hand, the Arnøytind and the Skartind, purchased after the crisis, may be thought of as part of Arnøy's post-crisis 'restructuring.' Alongside these relatively large vessels, are the many smaller vessels, the common thread that unites the fisheries of 1995 and 1997 with those of a generation or more ago.

The small boat fishery is a symbol of the viability of fishing communities. Although, as explained below, the numbers of small boat fishers have declined and this fishery is threatened, it is important to note that many men, including those outside the fishery, expressed the *ideal* of owning their own small boat, an ideal with deep roots in North Norwegian culture. In the words of one Troms county official, *for the fishermen it's attractive to have their own* ***sjark,*** *to be their own masters, they have a quota, they can take it when they want.* One Årviksand fisher, who spends much of his year fishing on large vessels far from home, justified his intention to someday buy a small boat, as it would allow *more time at home*. Thus, while larger fishing vessels may be the efficient economic future of the industry, small boats remain its heart and soul.

With a mixture of sentiment and experience, Arnøy people equate the future viability of coastal life to maintenance of the small boat fishery.

The epitome of the coast fleet is the *sjark,* a motorized vessel around 30 feet long that is usually fished by one or two fishers. A small boat fisher from Langfjord first answered my query about the length of his vessel, before describing his yearly fishing cycle: *nine meters. I fish by myself. I fish all year round. Right now, in December (1997), I fish for* ***sei*** *or* ***kveite****. I fish one and one half hours from here. After January, I go to Finnmark to fish for* ***torsk****. I go to Finnmark in February. I go by myself. I use net, gill net. February, March, April and early May. In May I fish with [long] line or* ***jukse*** *[jigger]. I stay a month, two weeks, three weeks. I sell my fish in Finnmark. In June and July I fish here with machine* ***jukse*** *[i.e., mechanical jigger], for* ***sei****. In August I fish too. I sell it in Årviksand and Arnøyhamn. I use long lines and fish for* ***hyse****. This would be near here. This continues to October. Also I fish* ***sei*** *with net.*

Another small boat fisher, from Haugnes, fishing alone from a 31 foot fibreglass boat, built in 1984, and purchased in 1986, told me in 1997 that he, like others, fishes squid (*akkar*) in the fall. He then heads to the Lofoten Islands in January to fish for cod using mechanical jiggers rather than nets. In his view the Lofoten fishery is over-crowded with nets. He claimed to be only one of three small boat fishers from Arnøya who still journeys to the Lofoten Islands, adding that there *aren't so many of us left*. The seas around Arnøya are too demanding in winter, whereas those around the Lofoten Islands (or in fjords) are more protected.

A third example of Arnøy's small boat fishers is that of two young Langfjord brothers who fished during the winter of 1995 from their 34 foot wooden *sjark*, the Havål. The brothers fished locally for cod, using baited long lines and keeping six locals in Akkarvik employed between January and April baiting their long line hooks. An Akkarvik man characterized the two brothers as real fish wolves or 'cowboys of the sea' (*ulver* meaning wolves, or in local dialect, *ulvan*). This revered characterization (similar to the respected 'fish killers' of the Newfoundland fishery) applauds the brothers' bold willingness to fish in any kind of weather, and their generally good catches.

Unfortunately, the Havål broke anchor during a winter storm in January 1996 and ran aground.[6] The brothers combined their own money, an insurance settlement, and a loan from the Statens Fiskar Bank and then approached Skjervøy *kommune*. The *kommune* government first promised a loan of 200,000 NOK, an amount subsequently reduced to 55,000 NOK. The two protested, questioning the sincerity of the *kommune's* 1996 commitment to help young fishers, and justifying

[6] My description of the brothers' ordeal in this paragraph is based on a *Nordlys* news story from 20 May 1997 (RL translation).

that they would be employing several people ashore, baiting long line hooks. Although the brothers eventually resumed fishing in a new *sjark*, their saga illustrates both the economic spin-offs of the small boat fishery, especially when fishers use baited long lines, and the risks inherent in the fishery, risks that ever fewer young people appear willing to take on.

It is important to note that in the first two cases, these highly skilled, hard working, successful small boat fishers claim they are in the last generation of their occupation. They may well be right, but if so, why? Arnøy people commonly cite four reasons why the small boat fishery is threatened. First, they argue, government and industry prefer larger fishing operations; second, efficiency within fish plants; third, possible changes in the behaviour of fish; and fourth, obstacles to recruitment of young fishers. This final 'threat' to the small boat fishery is part of a larger, looming labour shortage within the fishery.

On the issue of size, an Akkarvik fisher, who in 1995 fished for a quota of 15 tons of cod from his 22-foot boat, made the commonly heard claim that when the boat quota (IQ) system was introduced in 1990, quotas to small boat fishers were proportionally smaller than those to larger vessels. In his words, *when the crisis occurred the Norwegian government went too far, punishing many small boat fishers in the fjords. The ones that were fishing in the fjords were the ones hardest punished by government. When I fished in the fjords, I would never take that much [fish]. Boat quotas for small boats were proportionally smaller than those for big boats.* An Årviksand woman made a similar point, also tracing the problem to government's response to the crisis. *We have been fighting for keeping the small boats. Government says that* ***fabrikk*** *[factory] ships should have big quotas. We fight against this. The Norwegian* ***fiskarlaget*** *[union] opposed this [government policy]. Yet all statistics show that it's the small that survive and they [government] do not put their finger in the ground and know where they are. Yes, we've lost a lot of boats.*[7]

Secondly, fish plant managers reward larger vessels using the rationale of economics of scale and efficiency to show that larger catches make better use of time and labour than smaller ones. Managers seek continuous supplies of fish, preferably of the same or similar species, in order to efficiently operate and maintain an appropriate labour force. And, of course, the political clout of fish plants as employers of people, tied as they are to communities, hastens this trend toward larger scale. Large vessels regularly supplying large quantities of a limited number of species ensure predictability and efficiency, whereas the smaller and more diverse catches of small boats take more time to process, increasing processing costs. The manager of Arnøybruket put it this way: *with a small catch, here at Arnøybruket we weigh the whole catch, as they often have a variety of*

[7] Brox (1998) notes that although fishers would prefer small boats (under 30 feet), the vessel quota system favoured larger, more expensive vessels. He advocates quotas to fishers, rather than boats, so as to generate employment, yet he fears a political move toward ITQs.

species. Also the big boats can get a good price because they have a big catch. Its cheaper for us with big boats, it's more of an assembly line. This is one of the reasons why the plant likes big boats. The same manager illustrated the preference for large boats with examples of landing contracts between Russian trawlers and large Norwegian plants in Finnmark, enabled after a change in legislation in 1990 (or 1992).[8] He said, *big Norwegian plants have contracted from Russian boats; Norwegian [small] boats cannot regularly supply fish so as to keep the plant going.* In 1997, the young manager of Skarbruket discussed this issue with me. In his view, *the entire fishing industry favours bigness. Government contributes to this, with departments like SND favouring bigness. It is the economics of scale which works this way: If a small boat delivers 100 kilos of several species of fish to a plant, the fish must be sorted, weighed, processed, and packed, all of which takes the time of several workers. If a larger boat predictably delivers a 100,000 kilos, preferably of the same species, plants know how many people to employ and can employ them more economically.* In short, there's more predictability in large size. Large size reduces the costs per unit, removing uncertainties. This, the manager concluded, is one reason why small boats are leaving the fishery, with ominous consequences for community futures.

Thirdly, Arnøy people believe that ground fish species no longer behave as they did before the crisis. When accounting for why the 1995 winter cod fishery for small boat fishers near Arnøya was not particularly good, one former fisher explained that cod were *very deep and looking for food.* A Lauksletta man related this problem to small boat fishers. In his words, *as it is the fish don't come to the coast, so they [small boat fishers] get nothing and they have to live with this. Lately the fish is far away and it's very deep. They get no fish on* ***garn*** *[net], only on* ***line*** *[baited long line].* A plant worker and part-time fisher from Årviksand described changes in fish behaviour and the problems for small boat fishers, as follows: *there was some fish but not acting normal. Most of the fish did not do normal movements, to the Lofotens. But this winter [1995] they didn't do that; they just followed the food. The Alfredson was fishing near the* ***lodde*** *[caplin], and that's why they did well. The Alfredson could follow the fish, whereas the small boats could not do so, it was beyond their range.*

A final threat to the small boat fishery, and indeed, to much of the entire industry, is the recruitment of new fishers. Recruitment is less a problem on larger vessels, such as factory trawlers. Young people consistently told me that they would work on large, modern vessels but not in small boats. There are many obstacles to recruitment of young fishers: difficulties of obtaining one of few recruitment quotas; the high costs of acquiring a boat; the loneliness and discipline required to fish, and others.

---

[8] See footnote 19, this chapter.

Legislation requires that all children remain in school until age 15 or 16 but perhaps more important is the cultural change which increasingly considers education essential for most valued kinds of employment. As explained in Chapter 7, most Arnøy youth leave the island for secondary schooling and view such a career path preferable to work on a fishing boat. This cultural change is rooted in changes occurring after the Second World War and especially since the 1960s, when parents and teachers regularly advised youth to get an education. The following account, from a Lauksletta man, is typical of many accounts I heard: w*hen I grew up, parents told children to go to school, get an education, get a job, don't work with fish, that's the lowest [occupation] you can get. And that's why there's not so much recruitment. Why should you be out in bad weather when you can sit in a warm office?* Young people took the advice of their elders, many left Arnøya and the fishery forever (See Chapter 7).

The mayor of Troms Fylke, explained the broader context of the recruitment problem, concluding that youth increasingly pursue careers other than fishing. *I think that the parents do not motivate the young people to be fishermen. And then you have the cultural problems. In the 1960s, when I came to Arnøya, young boys lived in their harbour. They were familiar with the fisherman's culture. And there were a number of larger boats which they could take a trip with, and go ashore afterwards, or, if they wanted, in their fathers' boats. But we recognized during the 1970s and 1980s, that [for] the young boys, the life around the harbour was more removed from them.*[9] [Why?] *In these years the long line fishery was at a very low stage. Long line fisheries could occupy [employ] very many people: the women, children, and then they were drawn into the milieu, the environment of fishing, the culture, and they get in close contact with their fisherman fathers. So, in the 1960s, it was more a family-based fishery. The children were paid. Nowadays, the young men, the boys, are more connected to the schools, to the activities around the school, to bikes, and the like. I think that it's more a question of culture. In the 1960s, they were more comprised in a total community, a fishing community. I think (today) there's a growing [i.e., renewed] interest to be fishermen but today it's more closed. You have an arrangement with a recruitment quota but very few has possibilities to get such a quota. And we are now discussing how to mend that problem. Our [federal] Minister of Fisheries, Jan Henry T. Olsen, has a program allowing students to fish and sell fish in their summer vacation. There are opportunities in summer to catch* **sei** *and cod and perhaps young people could finance their school. [Also] we have built up some education for fishermen or young people who want to go into the fishing industry, and aquaculture. But there is no interest among the 16-year-old youth to go into these programs. We have in Skjervøy, in Tromsø and in Ebesøy, in the southern part [of Troms Fylkes], I think it's less than 15 [youth] from the whole county, age 16-19. The number of people*

---

[9] For a similar description of informal and formal socialization that formerly produced new generations of fishers, see Hetland (1984).

aged 16 to 19 is about 7000 in Troms. It's an alarming situation. It's a product of our politics, our cultural development, it's very alarming but we don't know how to cope with it. As I said, in the western part of Norway [i.e., the industrial fishery based in ports such as Ålesund], they do not appear to have the same problem. There's no recruiting problem. These are ships much larger than the Alfredson. But there is balance, politically and environmentally. We do not want to build up a large fleet of large trawlers -- ***Fabrikk*** *trawlers. This is now a national political decision. Norway doesn't want more* ***Fabrikk*** *trawlers. Because these trawlers only come into harbour to off-load fish for market, or, perhaps they go right to Denmark or England. There is much opposition to larger boats, especially in North Norway, initiated by Ottar Brox, and other scientists. Ottar Brox says that if we want to save our small communities, our dispersed population, it is necessary to have a coast fleet of small boats. They are most profitable. They are ecologically sound. There has been a trend toward more small boats, the* ***sjark,*** *but now we see the problems, especially in the northern part of Norway with having a fleet of only small boats. In Finnmark, this is a very large problem. In Troms we still have a fleet of larger vessels. It's a problem for our fish industry [i.e., plants], it's a problem for our markets, and a problem for recruitment. And we have to find the right balance between the small, the medium, and the large boats in every county. If we could find this balance, it [would be] necessary for us to have disposal rights over the resources -- to have control over the quotas. Then we could compose a structure of our fleet which would be most profitable in Troms. The proper balance of small, medium and large boats is a political issue. In our county plan, we say that the coast fleet, which means more the medium sized vessels (like the Alfredson), is the basis of our fishing fleet. And for us, it's very important to develop that fleet. It's able to move to seek the fish along the coast; it can give some comfort to the fishermen, and be attractive for the young people. But we have to have a trawler fleet too, but there's no political interest that the trawler fleet be increased in numbers or capacity, because of the resources and because of the total profitability. The fishermen aboard these large boats they have profitable work but the total profitability is not so good. Because there are not jobs in plants and because these large vessels are very, very expensive. They have to have a large part of the total resources to handle their cost. We also want a fleet of* ***sjark****.*

Data compiled by the Fishery Inspector in Skjervøy shows that in 1996 one half (36) of the seventy-two vessels on Arnøya were less than six metres, and that between 1990 and 1996, the number of vessels on Arnøya declined from 103 to 72.[10] The numbers of fishing vessels (large and small) have declined more dramatically over the past twenty-five years. An Akkarvik fisher claimed that in 1975, about 20 vessels of around 20 feet in length fished from that community; by 1995 there were only five (three belonging Langfjord fishers). And while the same

---

[10] Øyvind Isaksen (Personal communication 1997)

man professed optimism about the future of the small boat fishery, primarily because, he claimed, most of the overall IQ is allocated to small boats, my sense is that most would agree with a Langfjord fisher who predicted to me that *the small boat fishery will have a slow death. I learned from my father. But the culture around small boat fishing will die. It will not die fast but will be a slow death.*

Finally, another change dissuading men from becoming fishers has to do with contemporary expectations of uncertainty, predictability, to changing gender roles within families, and to fishers finding women willing to marry them.[11] Today, women expect their spouses to participate in the responsibilities of home and family, more than was the case in the historic, and highly gendered *fiskarbonde* economy, when no one expected that a man would wash dishes or help to get the children off to school. As we saw in Chapter 1, back then, men were away from their families for several months each year, during the seasonal Lofoten Islands and Finnmark fisheries. Wives back home accepted their responsibilities for the sheep barn and children. Much has changed today. Contemporary women are less willing to accept long absences from their spouses, absences still common to fishing. Women today want greater predictability and expect spouses to help with the daily tasks of home and family. These new expectations narrow the field of potential spouses for those wanting to be fishers.

Fish Plants in 1995

On Arnøya, as in all regions dependent on the fishery, there has been for some time an essential interdependence or symbiosis between fish plants and fishers.[12] Plant workers and fishers each bring different expectations to this relationship. Fish plants require fish, and when they have it, employ people. The local and regional economic significance of wages earned in plants explains the political priority -- both within the kommune and county -- of sustaining the plants themselves. Fishers, as producers, rather than processors, both those employed on large vessels and those owning smaller ones, are driven to catch the quotas assigned to their boats and bound, where possible, to land catches at ports where their wives, kin, or friends work. This symbiosis or embeddedness predates the crisis of 1989.[13]

---

[11] This paragraph is based primarily on conversations in 2000 with Helene Solvang, an Arnøyhamn woman educated in social sciences at Tromsø University. For a published account reporting increased male involvement in domestic responsibilities in the fishing communities of Finnmark, See Pettersen (2000:86).

[12] For more on this interdependency, see Jentoft and Wadel (1984) and Arbo and Hersoug (1997).

[13]This portrayal of relations between fishers and plant workers illustrates what Apostle et al (1998: 236ff) and Hovgaard (1998) mean by 'embeddedness,' a concept based on Mark Granovetter's (1985) reworking of Karl Polanyi's essentially functionalist notion of "interconnectivity" or integration between various groups and/or institutions. <u>If</u> modernity has disembedded economic arrangements and social relations (and this may be true in the general sense that Giddens (1990) claims), Apostle et al

The history of Arnøy's fish plants was discussed in Chapters 2 and 4. Here, I describe the ethnography of the plants as observed during fieldwork in 1995 and 1997. Arnøya has two fish plants: Arnøybruket and Skarbruket. Additional plants operating during my fieldwork near Arnøya include the Kræmer-owned Skjervøyfisk at Skjervøy, Johannes H. Giæver at Havnnes, and the Gamst plant at Rotsund (the latter two in Lyngenfjord). Arnøybruket (literally the Arnøy fish factory), replaced the previous company, Leinebø and Larsen at Årviksand in 1992 and is owned by Tromsø-based Norfra (literally meaning 'from the north'), a firm that originally specialized in exporting fish and in salmon aquaculture. In 1995, Norfra's managing director explained the company's corporate structure: *We have a board, with a chairman. We changed the structure of the company last year (1994) so that there is now a Norfra owning the different daughter companies: Arnøy-bruket, Torsvåg [fish plant], Norfra Export and our company in France, Norfra S and L. Norfra owns shares in the Skibotn [shipping terminal], as do the* ***kommune*** *and Norges* ***Råfiskarlag*** *[union].This reorganization in 1994, tried to give more autonomy to the daughter companies.* The managing director went on to describe Norfra's strategy for Arnøybruket, as follows: *When we came to Årviksand some 3 or 4 years ago, there were a lot of empty houses for sale, very cheaply, you could buy houses. This has changed, 90 degrees I would say. But we have been a bit reluctant to expand too quickly, we wanted to see that it is soundly based, and we wanted to do it slowly.*

The Arnøyhamn construction company of Sørensen & Sønn, and other companies renovated Arnøybruket and made substantial improvements to the wharf in the early 1990s, renewing the plant.

By 1995, Arnøybruket employed between 25-35 men and women, during the year between August and late June. The plant closes for holidays during July. Women tend to work on the filleting line, men in salting and transporting fish. During the field period, three local men worked as managers and another as a part-time accountant. Plant workers are unionized, represented by the Norwegian Food and Allied Workers (*Norsk Nærings-og Nytelsesmiddel-arbeiderforbund* [NNN]), with a local person as leader. In addition to national representation through the NNN, workers adopted a formal structure within the plant in March 1995. The plant

---

argue that to become sustainable, communities must somehow become re-embedded (as well as economically diversified or flexible), replacing individual self-interest with trust (no small feat!), thus generating social capital (See footnote number 2, in this chapter), and with it, enhanced sustainability. While sympathetic to their utopian prognosis, and as one also in search of community economic viability, I maintain that the task of measuring degrees of integration (or disintegration) in community contexts is more difficult. One might argue, for example, that relations between community and fish plant are more embedded in Arnøybruket than in Skarbruket, although whether this means that Årviksand is more apt to survive than Lauksletta is less certain. Consequently, the view taken here (see Chapter 8) is that no single explanation (such as social capital, embeddedness, or whatever) adequately explains community viability. Instead, a wider range of factors, empirically grounded, must be considered.

manager who, until then, resolved individual problems as they arose suggested creation of a formal structure. Consequently, in March 1995, workers selected one worker as *tillitsmann*, or shop steward, who now brings workers' concerns to the Manager. The shop steward explained some of the circumstances leading up to creation of this new, formal structure: *He [the plant manager] was having a lot of people in the office with problems so he told us it was time to select a person. So now if you are a worker with a problem, you tell me and I tell him [the plant manager]. We call it [the shop steward]* ***tillitsmann****, trusted man. I am just an ordinary worker. In the meetings, I am spokesman for the workers down in the salt [fish section of the plant]. I am listened to, I was elected in March, perhaps 5 years too late, to do better things for Arnøybruket.* Another worker holds the position as *verneombud*, investigating various health and safety concerns.

The plant management appeared to be progressive and fair-minded in its dealings with fishers and plant workers, though as explained below, Arnøybruket has its antagonists. On the positive side, however, in 1993 Arnøybruket's plant manager explored the possibility of a professionalization program enabling plant workers to upgrade individual skills and to enhance proficiency within the plant as a whole. The professionalization program aimed to meet new demands from European countries about the quality of fish (Guttormsen 1993b). The manager at Arnøybruket explained, ***Fagbrev****, is a program throughout Norway. I investigated doing it in 1993, to see if it can be done in evenings. They [workers] get more money but that's not the most important thing, it means they can work anywhere. We have four persons in Skjervøy and seven here [Årviksand] taking the course. The course started in 1993, with theory, and it continued in 1994, and we have some* ***praksis****, and we get a paper [diploma] if we do well.* One of the student-workers at Arnøybruket described the *Fagbrev* course as follows: *There is a lot such courses, in industry, in stores, and so on. This course we had is called Paragraph 20:* ***Fiskerindustri Fagbrev****. This is the contract I wrote with Arnøybruket. We take the course after work. This spring, four of us was having extra time because we* ***haven't*** *enough months of practice. We took another 5* ***fag*** *[subjects]: Norwegian, biology, business, data [computers], and environment. And rather than taking the Norwegian, we did the one [a local history course] with Helge [the course instructor].*[14] *Fagbrev* graduates receive a wage increase of 4.5 NOK per hour and, in principle, increased pride in their work. One *Fagbrev* student explained his motivations for taking the course: *for me the money (4.5 NOK more per hour) meant little. I took it to learn more about the fishery. It also gave me 10 points, half a year of post secondary education. I also got a book, and I've been into all chapters. I can do everything in Arnøybruket. So when we are finished with both the theoretical and practical parts of the course, people from Skjervøy will watch to see if we're good enough. No one has yet finished the course here.* Understandably,

[14] The Arnøybruket student-workers had the good fortune to conduct research under the direction of historian Helge Guttormsen, ultimately producing a report on the local fishery (Jørgensen et al 1995).

some of the student-workers said that some other workers at Arnøybruket resented the course, as explained by one student-worker, *a lot of the older people is working in the fishery think that all the new things, like courses, are unnecessary. They tried to tell us all the bad things about the course, like cutting the fish. When our teacher came down to show us, one of the older men stood close by and cut some fish. When the teacher told the old man he was cutting wrong, the old man was very sad. That is the teacher's view. When you are cutting a lot of fish, you need to hold the knife as the old man did. The teacher's way was the right way, but the heaviest way. When you're cutting a lot of fish, the old man's approach may be better. I learned a lot about bacteria, viruses, and so on. If we talk of something we learn in the course, the older people were always trying to sabotage what we learned.*

In 1995 and 1997, Arnøybruket had a verbal agreement that the Alfredson, of which it is part owner, would land fish in Årviksand. As well, Arnøybruket also purchases fish from numerous small boats. Fishers are guaranteed payment for fish delivered through a system[15] described by Arnøybruket's manager: *fishers always get their money. We pay* ***Norges Råfisklag*** *(the union) and they pay the fishermen. We pay [**Norges Råfisklag**] every week for the fish delivered that week. And they pay it out to every fisherman. Before 1990, we would pay each fisherman. But some plants had problems and they couldn't pay. But now you must have a bank guarantee for so many million, that you can pay fishermen.* [John: So, the fishermen are paid by the Union?] *Nearly [i.e., more or less]. But I must guarantee and every week I must pay in so much according to what I've bought (from fishermen), and the fishermen are guaranteed. If the plant goes down, the fishermen are paid.*

Arnøybruket produces several products: filleted ground fish, salted ground fish, and occasionally stockfish. However, the exportation of chilled raw cod, redfish, haddock, and other species to French markets is the plant's main commodity. Refrigerated trucks transport fish to Norfra's holding facility in Skibotn, Lyngenfjord, then through Finland, Sweden, Denmark, and Germany to France and Belgium, taking around three days. Codfish are priced and processed by size and quality. The plant manager explained, *the biggest codfish, say over 2.6 kilos goes to salt fish production. Salt fish markets pay more for large fish; it's the best price. It costs too much to fillet. Under 2.6 kilos goes to the fillet [line].* Markets for cod fillet are global, as the manager explained, *the fillet is frozen, it goes to England, USA, and now 20 tonnes goes to Iceland, and I hear they package it in smaller packages and they sell it to the USA. We have some problems this year (1995). We are getting 20 % less for the fillet we sell to the USA. The best price they pay is 35 dollars, now they pay 29, nearly 15 to 20 % [less]. So we see that it could be a minus for this year. And we have 150 tons of frozen fillet down in the freezer. In the USA, we deal with Boston. We use 2 middlemen: Frionor and Nordic*

[15] For a description of Norway's distinctive system of fish sales, developed in the 1930s, see Apostle et al (1998) and Trondsen et al (2003).

Netting lumpfish off Årviksand. 1995. (Photo by the author)

Group, they go and visit the customers. They set some rules for us how to produce, and we use their name on the pack. In the USA, the prices of beef and chicken affect fish prices. Arnøybruket has a mechanical cod splitter, which splits large cod for salting. Fish by-products, such as bones and cod tongues are sold, with the cod bones going to a fox farm in Finnmark. Because of variable market conditions, Arnøybruket stockpiles salted cod, as was the case in June 1996.[16] All of this salted fish is dried in commercial driers at Ålesund, west Norway.

Although everyone agrees that Arnøybruket is vital to the Årviksand (and, to the broader Arnøy) economy, the plant and parent company have not escaped local criticism. Four criticisms are most common: low salaries, worker allegations that the company ignores their opinions and concerns, low morale, and the plant's relative lack of secondary processing. One outspoken worker complained about wages, *we are having [had] a meeting with them [management] this spring, before the plant closed for holiday, and we asked for a better salary, I checked it out with five other plants and we are 9 NOK lower than what they pay their workers. And I have told the leader [manager] of Arnøybruket and we asked for 5 NOK. 'No, they couldn't'. I asked for 3 NOK. No again. So, from now on, every worker is very pissed off. Perhaps they couldn't give us this money, I don't know. Not all the workers asked for 3 NOK. I asked for my group, people working down in the salt, who go in the trawler to pick up fish, not for the women [such as on the filleting line].*

[16] A story in the Skjervøy newspaper, *Framtid i Nord*, on 6 June 1996 stated that Arnøybruket had 400 tons of salt cod in storage, and a work force of 25 workers.

In addition to alleged low worker salaries, one Arnøy fisher complained that Arnøybruket paid fishers less for fish than did Skarbruket, in Lauksletta. In his words, *that is because of local owners [at Skarbruket], they [Arnøybruket] could if they want pay more but they only pays list price. The list price is the least amount of money they can pay me. They [Arnøybruket] can pay more.*

Several workers at Arnøybruket claimed the company did not respect its workers: *perhaps we are good workers but they don't see our quality. They want us to be machines.* The same worker continued, *there are a lot of problems with Arnøybruket. There are a lot of [salt] fish they haven't sold. And we have some meetings, they owe the banks 1 million NOK. There are problems with management. It's hard to explain. We tried to have a formal meeting, earlier there were just informal meetings. It's the way they treat us, the ordinary workers. It's not just the local leaders; it's also the leaders in Tromsø. They don't know what's going on in the fishing industry, in the factory. They can just sit down with the calculators and see totals. They don't know what's going on in factories. They should come when we have a lot of fish and see how we work.*

Another worker, an ingenious plant engineer, devised (initially with the encouragement of the plant manager) a method of heating local houses by using waste heat from the plant's freezer engines but was disillusioned when Norfra head office failed to implement his method. The engineer explained the method to me, as follows: *The heat is coming from the freezing engines. We have one freezer today; we use 91 kilowatts. We have a new engine. We cannot use it until the electrician arrives. It's 140 kilowatts. When both engines work, you can produce 300 kilowatts. Now the heat goes to the sea.* [John: how does the heat leave the building?] *The water comes from outside and heats the gas, from the engine, ammonia. Heat comes from condensation. The water is heated from the gas to a temperature of about 9° centigrade. That was with the smaller engine. It was minus 8° outside. So now, we expel heated water into the ocean. There is presently enough hot water going into the ocean to heat 60 houses. If a house uses 5 kilowatts to heat, we could heat 60 houses. I spoke with the technical people in Tromsø about this. All this would cost about 200,000 NOK to harness this heat. This would be 1,752,000 kilowatts per year, with a savings of 613,200 NOK savings. [The plant manager] encouraged me on this but nothing came of it.*

Worker perceptions that Arnøybruket ignores their concerns may help to explain a survey conducted by Nord Troms Bedriftshelsetjeneste around the time of my 1995 fieldwork. According to Arnøybruket's shop steward, the Bedriftshelsetjenesten's service is paid for by companies such as Norfra and investigates worker absenteeism and related morale problems. The shop steward explained that a questionnaire conducted by the Bedriftshelsetjenesten found that: *thirty-eight percent of people did not like to go to work at Arnøybruket. That tells us something. This finding was based on a lot of questions. There are a lot of problems with this work. It's hard work, cold and monotonous. Another man from the fishing industry told us that 10 percent was always sick. Workers don't stop in their work to do exercises to warm up their body. If you don't have leadership who understands these problems, then problems continue. Fish cutters must work very hard. I think the most serious problem is the way the leader [plant manager] cannot catch the attention of the workers. The workers don't want to do their best and don't give a damn about the fiskebruket. Now, [the plant manager] is a very good man, but you know there are others under him. Perhaps there's the problem, more than [the plant manager]. [The plant manager] is doing as much right as he can. Perhaps the Tromsø owners aren't giving enough power to [the plant manager]. They should let him do more. There is also a group of old men who have worked in the plant for a long time and don't like anything new.*

Although not discounting the importance of criticisms such as low wages or alleged company disregard for worker initiatives, perhaps the most vital critique of Norfra's Arnøybruket is, like small fish plants elsewhere, that most fish leaves the community unprocessed. Ironically, unlike in the past, when stockfish left North Norway as a value-added product, there is now relatively little adding of value (or secondary processing). Excepting production of wet salted fish, exported to West Norway to be dried, value is not locally added to raw fish. I asked Norfra's managing director in Tromsø whether more value could be added to fish which now leaves Årviksand round, only to have value added elsewhere. He responded, *if you're going to make food, which can be heated and served, you need connections with very large companies, and to be part of their concept. Then you can be sure that you're able to deliver continuous supply. It's a problem in our area with a coast-based fleet. The companies own their own trawlers. They can compose their product. [Moreover, our] salt fish traditionally goes to lands like Spain or Portugal, as* ***bakalau****. But these round fresh fish goes to well-paying markets of northern Europe, Belgium, France, and so on. What about value adding? There is an objective to increase the value adding in our own communities. Politically it is popular. It seems to be very difficult. You see it in salmon. Almost all leave round. And the French and Germans smoke it. In LLFISK [a* ***fylkes*** *fishery council], we had a discussion about this fresh fish export and how to create more work in the communities. There is a culture in the market in some parts of Europe; they want to see the whole fish, with heads and tails. And you go in Brussels and you see whole fish. People have to see.* Norfra's managing director's argument that Arnøybruket is too small, and that the company's round fish are what Europeans consumers want may be correct, but one Årviksand man mentioned a small fish company in Skjervøy which adds value to local fish. In this man's words: *the company is called Nordtro. They make* ***fiskekaker*** *[fish cakes or burgers]. They employ peoples with minor physical handicaps. It used to make nets. It employs about 7-8. They also make salmon, crab burgers. It's an example of a small company.* [17]

---

[17] One suspects that subsidies might be available for companies such as Nordtro, given their social mandate to provide employment to handicapped persons, and thus reducing the validity of the man's

In any event, the goal of adding value to fish is very important to the people of Arnøya. Typical of local opinions on this are those of one man, who claimed; *I heard that they [Årviksand people] are not happy with Norfra because it was originally a pure export company. In Årviksand, that is why the fillet production is almost nothing. They take the fish over the quay and export it fresh, drive it to Skibotn and load it over to containers [with a] loss of filleting jobs.* Similarly, a Skjervøy man said, *they are delivering fish without working on it. Delivering round fish to Norfra. Working with this [adding value] could give more status to the fishermen.* Another person, a fisher, concluded more cynically: *But there's another side to that problem. There are now lots of owners of Arnøybruket. And they don't want Årviksand [people] to earn very much money. They will [want] that the export firm in Tromsø should earn the money and Arnøybruket should just go around [continue]. The money earned is in Tromsø, in exporting. That is Norfra's strategy. They pay very bad to the fishers and buy the fish from Årviksand cheap and sell it and the money is back in Tromsø. [There is the same policy] on Skjervøy, with Kræmer. They own boats. They buy the fish for the lowest price and produce on Skjervøy; the earning is back in Tromsø, in exporting. That's because it is not local owners in the [fishing] industry. That is a big problem. The only one industry that has local owners is Skaret.* This last person's argument concludes that absentee ownership appropriates economic capital from community, at the expense of the latter.[18] The broader regulatory context of this change is the deregulation of exporting regulations, which began in 1990, as mentioned in Chapter 4. Neoliberal deregulation eliminated middlemen, increasing both the risks and profits of exporters such as Norfra. However, these and other criticisms of Arnøybruket need to be kept in perspective: most locals believe that the present company, while not perfect, is far better than the Leinebø and Larsen company which preceded it.

As mentioned below in relationship to Russian trawlers landing fish in North Norway, and as explored more in Chapter 6, the main weakness of Arnøybruket is Årviksand's shallow harbour, enclosed by the man-made stone breakwater (or *molo*), described in Chapter 2. Some combination of tidal conditions and of the Årvik *elva* (the Årvik river), which empties into the ocean near the *molo*, means that the man-made harbour continually silts in. Historically, Årviksand's proximity to fishing grounds led to construction of the *molo* beginning in the 1930s and increasingly since, *kommune* politicians have considered Årviksand as the island's main fishing community, often at the expense of other villages. Of course the prime

---

comparison. Further, there appears to be some confusion surrounding Nordtro, some suggesting the company may have been in Nordreisa, not in Skjervøy.

[18] Arbo and Hersoug (1997: 122) identify the transition from local to multinational control of the Finnmark fishery as one of four forces of globalization of the industry. Lindkvist (2001) examines the possibilities for local development within the context of foreign-owned fish processing plants in Finnmark.

example of this political favoritism, discussed earlier, is Akkarvik, which ironically has Arnøy's best natural harbour. Lauksletta, or more specifically Skaret on Lauksund, has until recently, received less priority from *kommune* politicians, even though it too enjoys a good natural harbour and closer proximity to ferry services than Årviksand. Skarbruket, owned by Skaretfisk, is then, quite literally, Arnøy's second fish plant. A cynical observer might conclude that the poorer one's natural harbour, the greater the political capital it attracts but such a conclusion simplifies a much broader range of political decisions involving Årviksand.

Skarbruket, the island's second fish plant, is smaller than Arnøybruket and has a more seasonally variable labour force. In 1997, Skarbruket employed five workers in the autumn and around 25 during the summer. The plant is locally owned, started by a local man in the mid 1960's. Skarbruket is situated on Lauksund, the deep water strait between the islands of Arnøya and Laukøya, enabling larger vessels to land fish at the plant. Notwithstanding such natural capital, Skarbruket was less active than Arnøybruket during my time on Arnøya in 1995, partially, because of its reputation as an unpopular place to work, and because *kommune* politicians, especially Mayor Kurt Leif Strøm, originally from Årviksand, favoured his home community.

However, one of the changes I observed when I returned to Arnøya in 1997 was the resurgence of Skarbruket, as discussed by one person from Arnøyhamn: *you asked when you came what changes in the past two years [i.e., 1995 to 1997]. There is a more positive attitude in the districts [such as Arnøya] since Roy Waage has been mayor. When Kurt Leif Strøm was mayor, [he thought that] there should only be one fish plant other than Skjervøy on Arnøya and he did whatever he could to prevent Skarbruket to get a filleting line. Today, there are more positive minds to find a solution for Skarbruket and today Skarbruket is on the way up. And that is a very good thing. You can't get people in Lauksletta to daily work in Årvik. So you have to have at least one industry [fish plant] at each side of the island, to make the society function.* The renewal of Skarbruket began during the spring of 1996 when the plant's owner questioned the *kommune's* position on the plant.[19] Long time Skarbruket owner/manager Rolf Pedersen wrote the *kommune* government in April 1996, asking what plans they had for Skaretfisk. The letter prompted a May 1996 meeting, held at Skaretfisk, between *kommune* and plant officials. Pedersen underscored the economic importance of the plant: that it paid its workers 1,750,000 NOK in salaries, that the 12 or 13 workers were from the *kommune*, that the plant produces 70 percent salt and 30 percent fresh fish, and so on. Pedersen explained the possibility of a dramatic increase in salt fish production but added that to do so would require financial backing; otherwise, Skarbruket could devolve into a place where fish was only landed, but not processed. Although Skjervøy's new mayor (Roy Waage) did not promise Skarbruket money, he offered political

[19] Details about Skarbruket presented in this paragraph come from a 30 May 1996 story in *Framtid i Nord* (RL translation)

backing in Skaretfisk's attempts to obtain funds from elsewhere, particularly from SND.

In January 1997, Rolf Pedersen's youngest son took over management of Skarbruket from his father. The son described his background in the fishery to me as follows: *I am 25 years old. I learned on the job. I was with Hermann (Pettersen) in Portugal in 1994 and he taught me. And his company in Møre as well. I learned on the job. All these guys [in the fishery] want to help me, because I am so young. They think it's funny. I took over from my father on 1 January 1997.* In 1996 Skaretfisk produced about 8 or 9 hundred tons, and in 1997, two hundred tons. The plant mainly produces salt sei and torsk, because, as the young manager explained, *a lot of big companies produce fillets, we cannot compete with them so we produce salt fish.* Most of the saithe goes to Brazil; the cod to Portugal. The autumn season -- August to Christmas -- is the plant's slow season, as Skarbruket does not buy herring. Autumn is also slow because cod is in high demand but Skarbruket is unable to compete with larger fish companies that contract with the Russians who regularly produce large amounts, thus earning good prices. In all this, smaller fish plants and boats are less able to compete. However, after Christmas, Skarbruket's importance to the local fishery improves, as there is more cod, and better markets.

In November 1997, the young plant manager described their operation as follows: *right now we are buying from Skjervøy and Arnøy fishers. We have a coast fleet we are buying from and helping them in wintertime. We are buying from 12 or 13 boats. I talk with the boats during the season, when they are fishing elsewhere, and say if I'll pay the same, why not sell to me? The problem here is one of supply of fish and small companies like Skaretfisk face difficulties during the period between August and December when most locals have fished up their cod quotas and are turned to* ***sild*** *[herring]. Yet Skaret does not buy* ***sild****, and needs* ***torsk*** *during this fall season. We have problems competing with big companies. The five people working here are from Arnøyhamn. Our maximum labour force, during summer, is around 25, producing* ***sei.*** *We produce approximately 100 tons each week.* ***Sei*** *is difficult to process as we buy it with head on, guts in. You have maybe 12 hours to butcher it; if not, it goes bad. So, when we are processing* ***sei,*** *maybe 100 tons, we [process] 30 tons of [****dyrefor****, animal food] guts and head, that we need to freeze for sale to mink farms in Finnmark. We receive about 0.5 NOK for this. But it is difficult. It is difficult to produce [****sei****] with the bone out. That's what is difficult to get the bone out of small* ***sei.*** *This is between May and August. Now we have a [production] season of 8-9 months. But now with the new collecting system [see below], we're hoping to get another 2 or 3 hundred tons.*

In order to increase their supplies of raw fish during the slow fall season, in 1997, Skarbruket planned to truck fish from several small communities, including Akkarvik, to the plant. It is important to emphasize that as potentially beneficial to Skarbruket and communities as this proposed system is, it may not prove sufficiently profitable for Skarbruket to continue the system. Significantly, in 1997

Arnøybruket attempted to truck some fish from Djupvik (Lyngen Fjord) but stopped because the subsidy *Norges Råfisklag* gave to the plant was insufficient to show profit. Even so, for Skarbruket, the new collecting system, as explained by the young plant manager, appeared to provide his plant with a supply of fish during the slow autumn season, and provide a market for fishers in small communities. *This is a new system in North Norway. This system has existed [elsewhere] for a while but has not worked in North Troms because of turnovers in buyers. The Råfisklag pays me 1.6 NOK each kilo, for me to go with truck to get all this fish. This is to help the fishers [stay in small places]. We have some small [presumably chilling] facilities around and I am trying to get [this system] rolling, to fill the gap between August and December. Trucks will bring us the fish from Djupvik, Sørkjosen, Badderen, Burfjord, Segelvik, and Akkarvik. In all these small places, the problem is that all want some [buying system] that works because some weeks they produce very small quantities. The buyers lose money. So I told them during summertime, I can help you but I need everybody. You see in these places, they produce about 100 to 150 tons each. So I will have fish, between August and December and [I will pay] the established price list, nothing more -- not the plus 10 or more per cent [paid to fishers delivering at Skaret]. I own the truck, so the plant receives 1.6 NOK per kilo. And in the small places they know that they have someone collecting the fish. Not every time is there fish in each place. And the fishers know, I'll deliver fish there [at Skaret] because he helped us at our place. I can go pick up 200 kilos at one place. The fishers from other places without fish on that day don't worry because they know we'll collect from them when they have fish. Right now we've only began collecting from one of the communities, twice a week. The fishers store fish in the* ***mottak*** *[depot, in this case, for fish] and my driver picks it up. The fish is sold head off, but not gutted, kept on ice. I pay fishers 0.35 NOK each kilo. So if a fisher sells me 1000 kilos, I pay him 350 NOK because I am renting the* ***mottaks*** *the fishers own. It used to be that the communities had a man who worked in the* ***mottak****, he received 0.5 for each kilo that he packed. But I don't use the [local] man to pack and now use the truck driver to write the paper [invoice]. So I earn 0.5 NOK more. In the end that will work. Because if I have weeks where a place only produces 1000 kilos of fish per week, and I had to pay to have the man there, it wouldn't work. When we do it my way, it should work. But some fishers don't like it; because they have people they know who wants that job. But I had to tell them we must do it my way. Altogether, there are about 100 boats. I think it may be more.*

Skarbruket's plans for this new, decentralized system, as of 1997, potentially helped fishers and small communities survive. The manager of Skarbruket characterized Akkarvik, the only community on Arnøya under this planned collection system as follows: *Akkarvik is a very good place to fish from during the season. Akkarvik has 3 or 4 boats, but are joined by others from Skjervøy during the season. I think there will be about 150 tons at Akkarvik. But you know these small boats, now there's a new quota, that the small boats will have 600 tons, they will lose their quota if they don't catch their quota. We will drive a truck to pick up the fish.* In 1997, Skarbruket's proposed collection system, fish supplied by Russian boats, and a new, younger manager, promised to increased the company's prospects, thereby improving possibilities for people living on the eastern side of the island, from Arnøyhamn to Lauksletta.

Russian Fish for Norwegian Plants

In 1990 or 1992,[20] following Norway's short cod moratorium (1989-1990), a change in legislation permitted fish-starved Norwegian fish plants to land and process foreign (mainly Russian) fish. Fish processed in North Norwegian plants and sold as 'Product of Norway' helped Norway's export economy while at the local level, it preserved processing jobs following the crisis. However, dependence soon replaced supply. By 1992, Apostle et al observe that "50 per cent of total landings in Finnmark were Russian" (Ibid), and by 1995, the percentage of Russian landings rose to 52 per cent (Arbo and Hersoug 1997. 126).

There are three points I wish to make about this new reliance on Russian fish. The first briefly examines the political economy of North Norwegian fish plants, again primarily those in Finnmark, and questions who actually benefits from importing Russian fish. My second point questions the consequences of importing raw fish on Norway's inshore fleet, and is rooted in and illustrated by events occurring during the Spring of 1995, when Norwegian inshore fishers claimed they were turned away from Finnmark fish plants busy landing Russian cod. My third point returns us to Arnøya, where one plant (Skarbruket) lands some Russian fish and the other (Arnøybruket) cannot.

Beginning in the 1960s, many small, locally (even cooperatively) owned fish plants began to be replaced by larger plants, often owned by persons and capital 'outside' local communities. This change was associated with the decline of locally owned cooperative fish plants that arose during the 1950s and with the transition from stockfish to frozen fillet. Apostle et al write that "southern Norwegian investors, particularly from Bergen and Ålesund, became heavily involved in Finnmark ... " (p. 253) severing the embedded social networks that previously linked the plant to the community and fostering a level of dependent anxiety about distant plans for local fish plants.[21] Arbo and Hersoug report that by the mid-1990s, two-thirds of plants employing more than ten people were owned or controlled by shareholders outside Finnmark; that "12 of the 17 large trawler licences in

[20] Some confusion appears to surround which legislation was changed, and when foreign fish could be landed. Arbo and Hersoug (1997: 123) claim that in 1992 the Fishing Limit Act allowed foreign landings, while Apostle et al (1998: 231) date Russian landings to 1990, when the Sea Border Act was changed.

[21] Symes (2000:4) observes that in Europe "a weakening of traditional links between the fishing boat, the homeport and the local fishing grounds" challenges the concept of the fishing community; see also Hansen (1997) and the Epilogue, below.

Finnmark are controlled by Findus [Nestlé's huge plant at Hammerfest] and the Ålesund-based West Fish Aarsæther" company; that the ownership turnover accompanied both a reduction in the labour force and a change in its ethnic composition; and finally, that between "30 and 50" per cent of plant workers are now immigrant (Tamil) workers (p.129).

Such details prompt the question of who actually benefits when North Norwegian plants purchase raw fish from Russian vessels? Although a thorough examination of this question is beyond my scope here, the corporate takeover of north Norwegian plants by powerful West Norwegian (or international) interests certainly makes them potential beneficiaries. I asked one of Arnøy's plant managers whether plants pay the same prices to Russians as to Norwegians. He answered: *the cost of production is less for Russian boats, which often have crews of 20 men. They [Russian fishers] receive small wages. The Norwegian plants pay the lowest possible price for fish, and act as 'agents' for the Russian suppliers, taking an extra 10- 15 per cent as a fee. Say, for example, that the fish was valued at11.60 NOK per kilo, the Norwegian plants would pay the Russians 8.60.* However, while interviewing the owner of one fish company that operates several fish plants in Troms county, I rephrased the question as follows: If a Russian trawler lands in Finnmark, or Troms, do the Russians get paid less for that fish than a Norwegian boat would? The owner replied, *no, I think it has been equalizing out but there are some harbour duties, some piloting costs, which the Norwegians don't have to pay and there is some marginal tops, but given that the industry [plant] knows that it will get a catch of some 30, 40, 60 tons, that is efficient, instead of landing 500 kilos. And the Russians have been catching a smaller fish and you know a smaller fish for the filleting industry is cheaper, because prices are based on size, and smaller fish are thus a cheaper raw material for the filleting industry [cf. Arbo and Hersoug 1997: 131]. On the other hand, the Norwegian fishers want to catch larger fish because their quotas are limited and they want to have the highest value for their quota.* In short, predictable supplies of cheaper Russian fish allow plants to run at maximum efficiency, and saved some plants in East Finnmark (Apostle et al, p. 252). The absentee owners and shareholders of Finnmark plants profit from Russian fish, and exports of white fish help Norway's trade balance.

It also appears that within the context of economically troubled contemporary Russia, the Russian fishing industry benefits from selling raw white fish in Norway. Speaking primarily from his experience in Finnmark, the General Manager of Norfra explained the advantages for Russians landing at home and in Norway: *what is interesting is that local [Finnmark] fishers are not against Russian landings in general, but they want it reduced so that smaller boats can land. It is very difficult to see what will happen but I think it's a question of how competitive the Norwegian industry will be, compared to other possibilities for commercial landings. For some, Norway is a large factory ship, lying just where it is. And the landings are coming. It is economically wise for the Russians to land in Norway.*

They get paid, whereas in Russia they only get small amounts and the bigger part of the catches are leaking to the officials. One will get 40 per cent, another 20 per cent, and if they want to export they have to pay another 40 per cent in export duties. So, they are catching outside the Norwegian coast and landing close by, and that is for them very economical and for the Norwegian industry economic as well. In early May 1995, more than 40 small Norwegian vessels blockaded the harbour at Kjøllefjord, Finnmark, protesting that the local plant was buying Russian fish yet turning their Norwegian boats away. Media reports claimed that Norwegian fishers blamed Fisheries Minister Olsen for complicity and interpreted his lack of support for their plight as retribution for North Norway's opposition to joining the EU in the referendum of November 1994. News reports of the blockade triggered my suspicion that the importation of Russian fish by the new, largely non-resident owners of plants in Finnmark may benefit capital and the state, quite possibly at the expense of North Norway's inshore fishers. My concern here is that increased reliance on Russian landings to supply domestic plants may lead to the further deterioration of Norway's already aging and threatened inshore fleet. Arbo and Hersoug (1997) cite extensive data (such as a shrinking and aging domestic small vessel fleet) before reaching a conclusion similar to mine. They write: "Non-local owners have a weaker adherence to the local vessels and a more limited responsibility towards the local communities. When raw fish can be bought wherever it is cheapest, local fishermen have no advantage" (p. 141).

On Arnøya, Skarbruket (Skaretfisk) at Lauksletta, has purchased some Russian fish since 1995. The plant did not purchase Russian fish during much of 1996 but in late May of that year entered a new agreement with a Murmansk (Russian) trawler to supply fish. Although Skarbruket continues to buy primarily from local boats, and its reliance on Russian fish is small compared to that of larger plants in Finnmark, that situation could change if Skjervøy kommune's inshore fleet continues to decline. Of course, there are also positive consequences, for example, a Lauksletta shop owner, who at the time (1995) also ran the post office, described the positive economic benefits of the Russian trawlers: *And just this year we have dealt with the Russian boats. We have had Russian boats every week. They're from Murmansk. It has been good for jobs in the plant, and when there is no work they draw unemployment. We have [sold] some* ***grønnsaker*** *(vegetables): onions, carrot, and some coffee machines. Once, we sold 40 coffee machines to the Russians. They buy chocolate, and in the post office they use Norwegian* ***krone*** *(NOK) to buy American dollars, totaling about one million dollars! I sold one million NOK and bought American dollars. They've been good for us.*

Arnøybruket, at Årviksand, cannot purchase Russian fish because, as one man put it, *of the harbour, it's too shallow*. I asked the same man, an Arnøybruket plant worker, whether Arnøybruket or Skarbruket had the brightest future. He said, *if you asked me which plant has the greater potential, I think its Skaret, because they can buy Russian fish. Because they can produce lowest price for the market. Markets are less interested in quality, more in price.*

In 1997, I learned about an incident that occurred at Arnøya during the spring of that year. In March and April 1997, a Russian vessel from Murmansk anchored off Arnøya and its crew of 25 began buying surplus cod livers from local fishers. However, locals believe that liver purchases were not the crews' only interest. A series of break-ins and thefts from Arnøy garages was blamed on the strangers, leaving angry victims. Although it may not be appropriate to conclude my discussion of Russian fish with this incident, the incident, along with more positive images of benefits to Finnmark plants, suggests that opinions continue to differ on whether the importation of Russian fish will benefit North Norway and its fishery in the long term.

Salmon and Aquaculture

Arnøya has a small commercial salmon fishery during summer and an aquaculture industry, although locals consider neither part of the fishery. Generally, fishers involved in salmon fishing do not fish commercially for other species, such as cod or sei. Even so, I include salmon here, because of the overall similarities between harvesting, processing, and marketing of salmon, and that of other fish species.

As explained in Chapter 1, the rights to fish for wild Atlantic salmon (*laks*) are rooted in hereditary ownership of *laksesett*, the inshore locations where salmon nets are set, as explained by one man, *my uncle owns very good parcels to catch salmon. You have to own the land. In Norway, if you own land along the ocean, you own into the water as far low tide. In the old days, they said, as far as you could ride a horse. If you have the tradition of catching salmon, you have first right to use it. If unused, anyone can fish from it. My ancestors owned some of the best salmon fishing places, going back to my father's father. From where you see the mountain going into the sea, from there for approximately 1/10 of a kilometer, they owned. My grandfather and his brother used to own it [them]. But today, we (our family) are 80 or 90 people: too big. After that, my uncle had it. Nowadays, we change every year. One year I have it, next year him. These places are named and numbered. They are called* ***laksesett*** *[berths]. Today, there are two ways to catch salmon. One is by* ***not*** *[seine], another by* ***garn.*** *It's highly regulated now. Until about 1972, people fished for salmon using baited* ***angels*** *[hooks], on hand lines, using* ***brisling*** *(sprat) for bait. They were very successful. This fishing method took place for about 10 years, but numbers of salmon collapsed. And the Danish fishers continued with this kind of fishery until about 5 years ago. The Norwegian fishers were very angry. And after that the Danish used floating [drift] nets. Norwegians also used them for a while. With drift nets, you open the net when the stream [tide] comes, the salmon become meshed, you then go along and take them out.*

Another man explained contemporary rules surrounding *laksesett* on Arnøya. *If you own the 'ground' (you can lend grounds), you must register the must mark your berths, writing a number on the rocks, S 99, for example, everybody knows that that's his salmon.* Salmon nets must be at least 250 meters apart and similar regulations prevent fishers from placing nets too close to river mouths. Patrols of the *Kystvakta* (coast guard) monitor the short salmon fishery checking on what type of fishing gear is used where, when, and by whom. One salmon fisher explained regulations governing gear, *one cannot fish salmon from Friday to Monday. When you fish by net, you must only fish from Monday, 8:00 AM to Wednesday 8:00 AM; when you fish by* ***not*** *[seine], you can fish to Friday. A* ***not*** *may be set from 6:OO Monday evening till Friday. You can only fish for 10 days in July, only 10 days for the whole month. With* ***not*** *you can fish June and July, to the 5th of August.*

The dramatic decline in Atlantic salmon leading to the closure of commercial salmon fishing in most other North Atlantic domains will eventually end Norway's commercial salmon fishery.[22] The wane of wild salmon coincides with (and some believe is related to) the rise of farmed salmon. North Norwegians, such as at Altafjord (in Finnmark, north of Arnøya), also farm cod. Registered fishers catch small live cod in a *rundfisk* trawl, which they then sell to fish farmers who feed them in cages. The cod aquaculture industry appears shrouded in secrecy, because, Arnøy people claim, cod farmers guard trade secrets about how they acquire and raise young cod.[23] In the words of one Arnøy man, *they [cod farmers] don't want people sneaking around. That is my feeling. I think I'm right. The government can't control numbers of fish reported as young.* In 1995 two young Arnøy men discussed starting a cod farm, but to date the island has none. However, in 1995, Arnøya had two salmon farms, one at Lauksletta and another at Akkarvik.[24]

A previous salmon farm was started at Akkarvik around 1970 under the name Akkarvik Fiskeoppdrett. Leinebø and Larsen owned this farm during the 1980s and it was later one of six salmon farms owned by the Fiskegruppen, until the onset of the financial crisis in 1988, described in Chapter 4. By 1995, the Akkarvik salmon farm was managed by Felleskjøpet, an agricultural cooperative, yet by the time of my return to Arnøya in 1997, the fish farm had been purchased by Dåfjord gruppen and moved (temporarily) from Akkarvik.

Over at Lauksletta, in 1984 two brothers and a partner from Harstad started a salmon farm called Arnøylaks. Arnøylaks originally had a licence to produce 150 tons of farmed salmon. In 1988, the fish farms at Akkarvik and Lauksletta, along

---

[22] Drawing on research in two areas of coastal Norway, Jakobsen (1996) shows how strong competition during the 1990s led to the successful development of local capital in fish farming, despite increasing global competition.

[23] Through Arnøy contacts, I attempted (unsuccessfully) to interview people associated with a cod farm in Finnmark. Arnøy people interpreted reluctance of cod farmers to speak with me to secrecy that they claim shrouds that industry.

[24] There is a third salmon farm, on the adjacent island of Laukøy, called Nikkebylaks, and owned by Aksel Hansen.

Arnøylaks, Lauksletta. 1997. Photo by the author

with others in Norway, suffered losses because of the financial crisis following *yappitid*, as one man explained, *fish farmers used the bank called the Norion Bank. That bank went bankrupt [in July 1988] and fish farmers banking through them lost their money. This occurred throughout Norway. On Arnøya it affected the Akkarvik fish farm, and the one at Nikkeby [Nikkebylaks]. The problem for the salmon farms was that they had sold a lot of salmon right before the crisis and they didn't get paid for many tons of salmon. That was their problem. They froze the salmon to wait for better prices that never came.* Essentially then, when the Norion Bank called in its loans previously made to salmon farmers, it was very difficult for salmon farmers to refinance with other banks, with obvious results.

Both salmon farms continued after the July 1988 bankruptcy of the Norion bank (even though the Akkarvik farm appears to have changed ownership) and by 1995, both were operative. That year, Akkarvik Fisk A/S employed three people, who tended the farm's eight *merder* (salmon rings). The rings have computerized automatic feeding controls. The previous year (1994), Akkarvik Fisk employed four or five men and produced around 420,000 kilos of salmon (calculated on 8 rings, each with 15,000 fish, averaging between 3-4 kilos).

Arnøylaks, at Lauksletta, employed four people full-time in 1995 and one of the owners estimated that they would produce between 400,000 and 800,000 kilos. Much to the chagrin of one of the owners and as mentioned above, several of Arnøylaks' permanent workers moved from Lauklsetta to Arnøyhamn, as one owner described: *I am the only owner who works in the farm. Three others work with me, all from Lauksletta, but all have moved to Arnøyhamn after they married.*

Unlike the Akkarvik farm, however, Arnøylaks slaughters its own salmon (and those from Nikkebylaks), seasonally employing six to eight people for around one hundred days each year. One of the owners of Arnøylaks characterized this seasonal work force as follows: *These are housewives, some work at Skarfisket from time to time. Many of these don't want full-time jobs. We just slaughtered 3 weeks ago [May 1995], now nothing until fall, then October to January, with January the big month. It's stressful to find labour. I cannot demand too much from my labour. I can call the office of unemployment. Persons on the unemployment list have to take work.* The same owner of Arnøylaks claimed that obtaining labour is not a problem: *It's no trouble to get labour. If you have work, you'll have labour. Very often we get calls from people wanting work. If we need more labour, we'll have a condition that you have to live in Lauksletta. Maybe it's the middle of the night and we need people, and they're in Arnøyhamn and unavailable.*

Aquaculture is continually developing new technologies, as one of the owners of Arnøylaks explained, *new computers, new software. We have a silo, a blowing system, which blows food for salmon through tubes. You have tubes in the sea for every cage. My idea is not to let the machine handle one hundred per cent. The machine cannot account for all variables, tide, sun, rain, wind, etc. The machine cannot see if the salmon have had enough. Many using such machines over feed. I prefer to let the machine handle eighty per cent of the feeding. Then I observe the rest. That way you don't let the machine control you; you have some control. What will come in some years is a sensor under the net. You'll feed until that sensor says no more. Perhaps some kind of sonar. That will come shortly.* Just as aquaculture technology is advanced, so the marketing of the product is global, reflecting both competition and different preferences of consumption. In the words of one Arnøylaks owner, *the Japanese want fat fish; French want leaner fish; so different markets ask for different fish. But you have to have more fish available to support your customers the whole year. Until now, we in North Norway produce in fall and early winter; fish farmers in South Norway produce beginning in February but this changing now. If I want to be competitive I must have fish all year.* Whether the salmon raised are fat or lean fish depends on their diet and on their amount of exercise in the habitat where they are raised. Farmed salmon raised in relatively calm, fjordal waters are more likely to be fatter than salmon raised in waters with strong currents (*strøm* or *straum),* where the fish must move with the current. The salmon cages of Arnøylaks are located in the swiftly flowing waters separating Laukøya and Arnøya, and this presents its own problems and possibilities. One of the owners of Arnøylaks described the problems associated with the 'stream' or current of Lauksund. *Our problem is to manage the stream. You have to use big weights to get the nets [cages] in place. We use concrete weights; suspended in the water... they are about 80 kilo in weight. And seaweed collects on the front of the net, making a wall. Our company is the first to have a sensor to take account of the tide. Tide is very important. The stream [tide] is strong in Lauksund. It flushes from*

Slaughtering salmon at Arnøylaks. 1997. Photo by the author.

big fjords. The stream also varies with the temperature. When the temperature is high, the tides are bigger. On the other hand, Lauksund's current resembles the river habitats of wild salmon, and this reduces the costs of production, in so far as Arnøylaks can use cheaper feeds which are higher in fat, as one owner explained, *Fat is the cheaper substance fed salmon. For us in Lauksletta who have (our salmon in) a very fast stream, we can use fatty foods. Thus, this reduces cost of production. And a fish that is swimming eats better than idle fish.*

Much of salmon produced at Arnøylaks is exported to France, as explained by one owner, *we fill a trailer with about 20 tons, then it goes to markets in Paris. It takes 3 days to get there; it's sold as fresh salmon. For freshness the most important thing is to cool the killed fish to near 0° C as soon as possible. If you pack warm fish in boxes then the ice will melt and you have too much water in the boxes. If you chill salmon immediately, ice it; you can have fresh salmon for 14 days.*

As positive as salmon aquaculture appears for Arnøy's future, the industry does not escape local criticism. One problem is pollution beneath the fish cages, as discussed by an owner of Arnøylaks, *if you (the salmon farmer) pollutes the area below the cages too fast it [salmon] can die, lack of oxygen, it can even explode because it produces methane gas. It doesn't produce an actual explosion. But it's not in my interest to pollute, but to produce the very best quality.*

Salmon farms are sometimes blamed for problems in other fisheries. The wife of one Arnøylaks owner admitted that, *we have heard people say that the cod doesn't taste like it did before establishing the [Arnøylaks] farm.* Understandably, her husband dismissed this charge, as follows: *The sea has always been free, common. If I see someone begin to earn money on common land, then we may have a disease called jealousy. So far, no fish farmer has disturbed (real) fishers. Those who do complain are older people who may have a boat for fishing for home consumption and may complain about fish farmers. But real fishers don't.* We should not be surprised that the owners of Arnøy's surviving salmon farm defend their growing business. On the other hand, perhaps the owners' claim that the farm has had few negative impacts is correct. Certainly, I encountered relatively few criticisms of Arnøylaks.

The fishery remains the lifeblood of Arnøya, even though the two ingredients necessary for it, fish and workers to catch and process them, are less predictable than they were a generation or more ago. By the time of my fieldwork, fish stocks had recovered from the crisis of 1989, although fish stocks will always be a concern. Regarding the labour necessary to catch and process fish, I sense that changing values and expectations increasingly makes it difficult to recruit new labour to work in boats and in plants.

In 1995, the mayor of Troms Fylke, himself originally from Skjervøy, and a person with extensive knowledge of local history, put the problem this way: *The next 10 years we will have other problems, recruiting problems in our fleet. The fisheries are in a way closed. You have your rights there and it's very difficult for young people to get in. It's another effect of our large number of small boats, compared with big boats. The owner of the small boat drives it himself, there is no place for another to get in. In the western part of Norway [such as Ålesund] they have no recruiting problems, because they have a fleet of larger boats, comfortable, and profitable for the fisherman. Larger boats allow young people to get a place there, to be familiar with the fisherman's life, and get into the culture. In our small communities, like Årviksand, and others, there are a lot of problems for young people to get into the fishery.*

Yet as I suspect the mayor would agree, simply adding more recruitment quotas does not solve problems of recruitment for new fishers. Because the contemporary fishery requires expensive vessels and gear, and is difficult and uncertain work, not all young people are willing to accept the risks inherent in entering the fishery. Meanwhile, like rural peoples everywhere, the young people of Arnøya are increasingly bombarded by the glitter and promise of distant ways of living, lifestyles that appear more attractive and contemporary than fishing from Arnøya. Many young people opt to leave the island, the topic of Chapter 7, but I turn next to another issue important to the survival and viability of the island's small communities: transportation.

CHAPTER 6

UNITY AND COMPETITION:

CHANGING TRANSPORTATION POLICIES[1]

Like fish plants and schools, roads, ferry connections, and tunnels are important components of the island's infrastructure, and, as argued in Chapter 8, vital to the future of rural regions such as Arnøya. As infrastructure, ferries and tunnels separate or connect space. Decisions regarding them, while important to the whole island, can unite or divide communities. I want to briefly discuss three examples of regional transportation policy affecting Arnøya, and the reactions of island people to them.

The first of these occurred prior to my fieldwork (although the consequences continue). This policy changed the ferry connections to and from Skjervøy, the *kommune* centre. The second occurred during my fieldwork, reducing ferry connections to and from Tromsø, the *fylkes* capital. The third example involves what might be called 'avalanche politics,' leading to a proposal to build a tunnel to connect Lauksletta and Arnøyhamn, thereby bypassing potentially dangerous winter avalanches along the coastal road connecting both places.

Historically, ferry and steamer travel to and from Arnøya were probably as much south to Tromsø, as they were to Skjervøy. Older people recalled 'hitching' rides on the steamer travelling to Tromsø, by literally flagging down the vessel at sea. Of course historically, all long distant travel was by boat, both between the island's villages and occasionally, to and from the island. Scheduled ferry connections between the island and Skjervøy are more recent, reflecting the increased importance of the *kommune* capital and the development of the regional road system following the war. As one man put it, *the ferryboat started in July [28th] of 1970, linking Skjervøy and Arnøya. Before that, a trip from Årviksand to Skjervøy took twelve to fourteen hours.* The scheduled ferry routes that began in 1970 connected Arnøya to Skjervøy via Nikkeby, on neighbouring Laukøy Island. Depending on winds, the ferry made three or four round trips daily, beginning in Skjervøy, sailing around the exposed headland of Engneset to Nikkeby, and then on to Skardet, on Arnøya (See Figure 2). The paved road around Arnøya opened in 1970, allowing automobiles and local buses to connect communities on the island

---

[1] An earlier version of this chapter was presented to the MOST CCPP Workshop entitled 'Whether, how and why regional policies are working in concert with coping strategies locally' at Joensuu/Huhmari, Finland in November 2000.

with the ferry at Skardet.[2] As early as 1971, before completion of the low bridge across Skattør sound (see below), one Lauksletta man worried about the dangers associated with having to travel around Engneset to Skjervøy, concerns which would later fuel the need for a ferry terminal on Kågen.

Although the Skjervøy ferry and the road across Arnøya dramatically improved transportation to and from island communities and Skjervøy Island, two developments on the neighbouring island of Kågen during the 1980s would soon involve Arnøya.

The first of these was the construction of a bridge across Skattør sound, linking the islands of Kågen and Skjervøy, completed in September 1971.[3] The second is Maursund tunnel, proposed in 1987 and opened in July 1991, connecting Kågen with the Nordreisa mainland. The bridge and tunnel connecting Kågen improved mainland road connections with Skjervøy; both developments signalled that policy makers in the kommune and fylkes governments preferred roads to ferries.

These new road links to Kågen placed the long ferry trip between Skjervøy and Arnøya under an increasingly critical lens. Would it not be better, policy makers in Skjervøy and Tromsø wondered, for the ferry to sail to Arnøya from some point on Kågen's new road network? The answer was 'yes' and by the late 1980s, two locations on Kågen -- Storstein and Vittnes -- were considered as potential sites for a new ferry terminal to Arnøya (See Figure 2).

All Arnøya supported a shorter ferry route to Skjervøy. In 1987, an Arnøyhamn man organized a lobby group to press for the shorter route. In his words: *our group was about 11 people, 3 from each village. I picked them. I picked them from different communities, political parties and genders.* Over the next couple of years, the lobby group lobbied for a ferry terminal on Kågen to replace the longer ferry trip to Skjervøy. The group fed media interest on the issue, and sent information to officials in Oslo and to their six representatives in the fylkes government in Tromsø.

In January 1991, the Arnøy lobby group received a letter from the transportation department in Tromsø confirming that work on a ferry quay on Kågen would begin in 1992. However, the letter failed to specify just where on Kågen the quay would be located. This uncertainty split the Arnøy lobby group, fuelling a debate between Lauksletta and the rest of Arnøya and prompting at least one person from Lauksletta to resign from the committee.

---

[2]The road between Arnøyhamn and Årviksand was passable as early as 1968. The broader, regional context of road construction on Arnøya, as mapped by Brox (1984: 156-57), shows the few roads in Troms in 1945 compared with the maze of roads in 1984.

[3]That bridges can dramatically affect peoples formerly separated by water and by so doing create new regions is seen in Berg et al (2000).

What I shall call the pro 'Vittnes faction', a group spearheaded by people from Arnøyhamn, hoped that if the new ferry quay was built at Vittnes, then another quay might be built just across from Kågsund on Arnøya near Arnøyhamn. Although not intentional, the implication of a Vittnes to Kågsund link would benefit all communities *except* Lauksletta, and this fractured the Arnøy lobby group, isolating Lauksletta.

For their part, Lauksletta people favoured another location on Kågen (Storstein), and for simplicity, I call them the ‘Storstein faction.’ If a new ferry terminal was built at Storstein, Lauksletta people claimed, then the ferry could continue to land on Arnøya at the old quay at Skaret, located at the southern end of Lauksletta. Keeping the quay at Skaret also facilitated serving Nikkeby, a short distance across Lauksund. Both the Vittnes and Storstein factions buttressed their preferences with reference to natural conditions. The Vittnes faction argued that strong northerly winds would make it difficult for ferries to dock at Storstein. Vittnes, they argued, is more sheltered from northern winds, and only 1.5 kilometres from Arnøya. Indeed, the proximity of Vittnes to Arnøya, coupled with the relatively calm waters of Kågsund Strait, had centuries ago led Kautokeino Sami to choose Vittnes as the crossing place for their reindeer herds when migrating to their summer pastures on Arnøya.[4] On the other hand, the Storstein faction argued that Vittnes is further from the new Skattør Bridge to Skjervøy and situated at the base of high mountains. Any road to Vittnes would be expensive to construct and threatened by avalanches.[5] The Storstein faction stressed the risk from avalanches on both islands of Kågen and Arnøya. On Arnøya, in the estimation of a person from Lauksletta, every winter between 30 and 40 avalanches, most of them small, tumble off mountains above Singla, occasionally blocking the road between Lauksletta and Arnøyhamn. Consequently, the Storstein faction in Lauksletta worried that if Vittnes was chosen, *they* might have to drive to a new ferry quay nearer Arnøyhamn, passing beneath Singla. Tensions between the two factions increased during 1992. One person in the Vittnes faction countered the Lauksletta fear of avalanches by saying: *but **we** have driven under those avalanches for 30 years but when they should have to drive under them, they make a lot of high voices.*

A land use map from Skjervøy Kommune (1991) covering the decade from 1992-2002 depicts two details relevant here. First, the map shows the ferry terminal originating at Vittnes, complete with a road leading to Skattøy Sound. Secondly,

---

[4]For an excellent photographic account of a Sami woman whose extended family group (*siida*) still herds their reindeer to and from Arnøya, see Kvaal (2001).

[5]The threat of avalanches is explained by the fact that roads in mountainous North Norway are usually built parallel to and just above shorelines and are, consequently, overlooked by mountains. The varying maritime winter climate produces avalanches, covering roads below and sometimes causing injury or death to travelers.

Local ferry at Lauksund. 2003. Photo by author.

the map shows the Sami reindeer route (*reintrekk*) across Kågsund, just west of Vittnes. One Årviksand man said that *the ferry quay on Kågen was to have been further north [i.e., at Vittnes] but [would have] interfered with Sami corrals.*

While the debate over Vittnes or Storstein raged on Arnøya, central policy makers chose Storstein and in 1992 construction began on a road leading to a new ferry quay there. The ribbon cutting ceremony opening the new Storstein ferry quay occurred on 3 June 1993. Four months later, however, northeast winds opened 'round two' of the debate. Persistent mid-October winds prevented the ferry from landing at exposed Storstein, prompting it to proceed to the old quay at Skjervøy and, in one case, to continue around the eastern side of Arnøya to Årviksand! Tromsø's major daily newspaper, *Nordlys* (15 October 1993), quoted the ferry captain as claiming that the decision to construct the terminal at Storstein was scandalous. Back on Arnøya, the October winds, plus growing complaints about poor bus connections with the new ferry schedule once again unified island communities (including Lauksletta) in condemnation of the new Storstein quay. A November 1993 meeting held at Arnøyhamn between bureaucrats, politicians, and local residents, broadcast live by Skjervøy radio, enabled all sides to express their views. Notable is that Nikkeby and Lauksletta people agreed with others on Arnøya that construction of a quay at Storstein had been a mistake. For their part, *kommune* and *fylkes* officials explained that renovations to the brand new Storstein quay would cost between 20-30 million NOK, and were out of the question at that time.

Over the next several years, the people of Arnøya learned to live with the decision to place the terminal at Storstein. As one of the Vittnes faction put it, *we have to accept that Storstein will be there for the future.* Coping with the decision was easy during good weather. Arnøy people travelled to and from Skjervøy in a few hours or, drove to Nordreisa or Tromsø to conduct business, trips that were a vast improvement over the original Skjervøy ferry route. However, during the winter of 1996-97, the Storstein ferry terminal was damaged by severe storms, refuelling demands for renovations to the quay. On Arnøya, the veneer of unity, which had united people of all communities in opposition to the Storstein quay, shattered early in 1997 with the rumour that a provisional ferry quay might be built at Arnøyhamn because avalanches sometimes prevented people from driving to the old quay at Skaret. By the summer of 1997, in the midst of a national election campaign, the minister of communication promised six million NOK to rebuild the Storstein quay so as to withstand northeast winds.

While the local ferry quay case led Arnøy communities to unite or to compete with each other, my second example, which began during the 1995 fieldwork, united all four-island communities. This second example concerns the fast boat (or *Hurtigbåten*) that runs between Skjervøy and Tromsø, with the issue being a reduction in the number of ports of call between these two destinations. Like the local ferry running from Storstein to Arnøya, a private Tromsø steamship company (*Dampskibsselskap*) operates the *Hurtigbåten* between Tromsø and Skjervøy. When I began fieldwork in early 1995, the Tromsø to Skjervøy ferry made a total of seven stops, including two on Arnøya, at Arnøyhamn and, on demand, at Akkarvik, and at nearby Nikkeby. However, during the winter of 1995, the steamship company decided to reduce the number of ports of call, from seven to three, leaving only the stop at Arnøyhamn to serve all Arnøya. Precise reasons for the steamship company's decision are unknown, but the company was probably either attempting to cut costs, responding to increasing reliance on road travel, or providing an express service for the business people that it considers its main clientele. However, people on Arnøya and other ports served by the *Hurtigbåt* considered the Tromsø ferry an essential public service, one not limited to business travellers.

Along with previous events such as the closing of the school and fish plant at Akkarvik, the threatened loss of ferry service appeared to drive another nail in the coffin of a dying community. Reactions in Akkarvik to the proposed loss ranged from resignation to anger. The steamship company's proposal caused all Arnøy communities to pull together. Indeed, during that winter, an Årviksand man used the *Hurtigbåt* issue to exemplify instances where all island communities unite. In his words, *all four villages on Arnøya as one, when necessary they all meet together. They [all* ***bygdelag****] met in February on the* ***Hurtigbåt*** *issue. They collected a petition, sent to Troms Fylke, sending an alternative transportation plan for Arnøya. Troms Fylke sent back a third plan, the one which will be implemented. The Fylke argues it's a business boat, not a service boat, as we see it.* Yet when I asked another Akkarvik man about efforts to reverse the steamship company's decision to stop serving his community, he sounded less optimistic. In his words, *it was impossible [to change] because the company was privately owned and some of that company's people were sitting in Tromsø. They did try in Skjervøy [i.e., the kommune government] to send letters, but it didn't work. Of course it's not* *the same when something that been there goes away, of course it's negative.* Despite the combined efforts of Arnøya's four *bygdelag* and those of the *kommune* government, the steamship company launched its reduced ferry schedule in June 1995.

But the issue did not die there. People from Nikkeby (on Laukøya) and from nearby Vorterøy island, supported by the people of Arnøya, lobbied through newspapers and government to restore service to Akkarvik, Nikkeby, and Vorterøya. Local people emphasized the inefficiencies of the changed route and by August 1995 their protest succeeded in reinstating fast boat service to Nikkeby, when passenger demand warranted. Limited service to Vorterøya was also later restored but Akkarvik was less fortunate, losing its ferry service.

My final example concerns Lauksletta's longstanding campaign to have a tunnel built to Arnøyhamn, thus avoiding the avalanches near Singla. Tunnels have long played an ingenious and important role in the Norwegian road network, and since the 1960s have been enabled by the generous revenues of the oil industry. Lauksletta people are well aware of the benefits of tunnels such as that linking Tromsø Island with the Tromsdalen mainland, or of the Maursund tunnel. They also realize that the opening of the road in 1970 had on the one hand permitted them to drive around Arnøya but on the other hand had, because of avalanches, isolated them.

As we have seen, Lauksletta has a strong economic base, including a salmon farm, an increasingly important fish plant, a general store, and more. However, since completion of the road, the avalanches off Singla Mountain (Singeltinden, elevation 742 m) threaten the village of Lauksletta every bit as much as they endanger cars. Lauksletta's youth leave, sometimes for Arnøyhamn, and this fosters some resentment and envy between the two communities. One Lauksletta person involved with the local salmon farm put it this way: *once young people leave and congregate elsewhere, in this case Arnøyhamn, they draw others. Young people need other young people. We're in a bad cycle here. I don't see a constructive end to the situation. World wide, computers take jobs. For me, I try to make the salmon production full-time; young people want full-time work. But the thing which disturbs me is that I have two young full-time workers from here and as soon as they got the job [with the salmon farm] they moved to Arnøyhamn. This community lost them.* [John: Why?] *They have a school and we have an avalanche problem. They don't want the children to pass by the avalanches between Singla and Haugen. Last winter there was one avalanche. One year more than 50. Nobody has been taken so far.*

This person's concerns about avalanches and population loss are typical, and help explain why Lauksletta people consider a tunnel to Arnøyhamn (specifically through Smellskardet to Grunnfjorddalen immediately north of Arnøyhamn) as the best solution to their problem. As in most such issues Lauksletta's informal council, or *bygdelag* continues to lobby the *kommune* and *fylkes* governments for the

Smellskardet Tunnel. However, progress has been slow, as one Lauksletta person explained: *government has told us that the tunnel through the mountain is not on their immediate plans. But I think if they see the plan long enough, they'll accept it. Some years ago, (about 10-15) the bus loaded with school children was just missed [by an avalanche]. Every winter there are about 30 or 40 avalanches.*

Lauksletta people continue to press for the tunnel but elsewhere on Arnøya, people believe that such a tunnel is unlikely to be built and could deplete funds available for other transportation projects. One 'solution' to the avalanche problem offered by some Arnøyhamn people is that the ferry travel between Lauksletta and Arnøyhamn, bypassing the avalanche area. One critic from another Arnøy community responded to my question about how much the Smellskardet tunnel might cost by saying: *Thirty, forty, fifty millions NOK for 50 to 80 persons. It will not be built. It will never be built. The tunnel will never be done. Some of them from Lauksletta see that but they don't say it, they are afraid to open their mouth.*

While Lauksletta people and others on Arnøya disagree on the tunnel, everyone agrees that the long-term viability of Arnøy communities requires improved transportation services. People stressed that local ferry traffic had increased since the opening of the problematic Storstein ferry quay and that improved transportation had not caused an exodus of Arnøy's population. Indeed, one man compared Arnøya to other islands within Skjervøy *kommune*, where, as he claimed, *no roads, ferries, [and] now 5 people [live there] in winter. Many say you make roads and ferries to make it easier for moving vans to haul the furniture away. [But] if you did not have the tunnel or ferry, if you had not the hope of them, the furniture would have been moved away a long time ago!*

These three examples illustrate how Arnøy communities cope, not only with regional transportation policies, but also with each other. As with my earlier discussions of fish plants and schools, transportation policies illustrate how cumulative chains of events enhance the prospects of some communities more than others. That change in transportation policy occasionally involves competition is not, of course, unique to Arnøya. Indeed, when neighbouring communities cease to compete, they may have also lost the will to continue. The case also confirms the importance of environmental realities -- winds and avalanches -- in local reactions to changing transportation policy and reminds us that such policies can both unite and divide people.

The next chapter reports on out-migration and concludes that improved services are essential in alluring some who leave to return.

CHAPTER 7

THE CLASS OF 1974-75

As throughout much of the rural world, young people are leaving the villages of Arnøya, most forever. The title of this chapter, and its focus on out-migration and attempts to reduce it, sprung from a question my wife raised one day in Årviksand about a photo that hung in the bedroom of the house we rented. The photo collage showed 71 smiling student faces from the Arnøyhamn Ungdomskole (Junior High School, grades 7-9) for the 1974-75 academic year. My wife wondered out loud: where are they now? Over the next several months, I connected names to the faces pictured in the photo and then carried the list of names with me around the island. I eventually traced the whereabouts and occupations of most of what I am here calling the 'Class of 1974-75.'

The short answer to my wife's question is that all except seven of these students left Arnøya permanently. This fact speaks volumes about the island's decreased population. However, it must be admitted at the outset that the class of 1974-75 was exceptional in at least two respects. First, school enrolment that year was larger than previous and subsequent enrolments and, secondly, out-migration was far less in the 1960s and in the decades that followed the 1970s. Yet indisputably, rural youth continue to leave Arnøya, begging the question of why and as well, the challenge of what, if anything, can keep youth home.[1] The problem of youth out-migration and more general population loss is considered by most Arnøy people and politicians in both the Skjervøy *kommune* and Troms county governments to be the greatest problem facing the region. I share their concern but repeat that out-migration is endemic to much of the rural world and that the new, global 'pulls' are increasingly cultural as well as economic, complicating local efforts to keep youth at home (see footnote 2).

This chapter begins with the Class of 1974-75 and then explains why youth began leaving in the 1970s. I contend that the causes of out-migration, while still to some extent economic, are increasingly also embedded in a collage of global images and ideas that challenge local life. Next, I review local attempts to curb emigration, suggesting that such attempts, while beneficial in many ways, are unlikely to halt the flow. Instead, I argue that efforts might better be directed at encouraging some (especially young families) who leave to return later in life.

From the perspective of eastern Canada, where unemployment fuels out-migration, out-migration from Arnøya surprised me, given the availability of well-

---

[1] Solvang (1997) notes that most out-migrants from Arnøya between 1980-97, and especially between 1980-90 were between the ages of 0-19, and that the proportion of the population over age 67 remained constant during the 17 year span of her study.

paying jobs on the island. Explanations of out-migration and of rural to urban migration usually conclude that people move for presumed 'better' jobs, even when they move *to* other resource-dependent communities.[2]

First, a few particulars about the Class of 1974-75: as implied above, what I, merely for convenience, have called the 'Class of 1974-75' was actually three school grades -- seven, eight, and nine -- from that year. The Arnøyhamn school photo includes 58 students from Arnøya.[3] Of those from Arnøya, 24 came from Årviksand, 15 from Lauksletta, 8 from Arnøyhamn, 6 from Akkarvik, 3 from Langfjord, and 2 from Haugnes. I consider the actual numbers of youngsters from these Arnøy communities significant in themselves, as they illustrate the relatively recent vitality of localities such as Lauksletta, which now have much smaller

[2] A relevant study by Mewett (1980) attributes high levels of out-migration from the Isle of Lewis (off the northwest coast of Scotland) to the poverty of the island's mixed crofting (farming) economy. In general, out-migration from Lewis is the result of an historical process of 'peripheralization' occurring since 1900, decreasing the number of island-based wage earning opportunities. The incorporation of Lewis into Britain fosters underdevelopment, resulting in a local economy that cannot support its population. Significantly, Lewis Islanders now see migration as a necessity. This normative belief that people 'need' to migrate for economic reasons, is common knowledge and serves to reinforce the decision made by each person who leaves the island. Conversely, Lewis islanders consider those who do not migrate 'failures.' Thus, migration conforms to both economic realities and to cultural expectations. Stockdale (2002) emphasizes the importance of family and social networks on out-migration from two districts of rural Scotland (one being North Lewis), and on choice of destination.

Several studies explore various dimensions of rural to urban migration in Norway, once again emphasizing the economic and/or educational motivations for migration. Vilhelm Aubert et al (1970), whose comprehensive (but sadly unpublished) study describes 'North Island', just south of Arnøya, discuss how whole families moved to Tromsø, after 'North Island' men began working on offshore fishing vessels based there (p. 196). This newer, offshore fishery replaced the earlier fiskarbonde economy back on 'North Island'; families apparently preferred the convenience and services of Tromsø to the relative isolation of 'North Island.' Aubert et al also discuss the 'native theory,' consistent with that reported by Mewett (1988), that the best young people leave rural communities (p. 185). Ringdal (1993) discusses motives in relation to the distance that migrants travel and the duration of their stays. Those who moved for long periods of time from rural to urban Norway improved employment opportunities and incomes, but not necessarily their housing conditions. Several studies, beginning with Nicholson (1975), observe that more young women migrate than young men. Hamilton and Otterstad's (1998) comparative study links female out-migration in Norway to both the small size of places women leave and to the lack of local economic opportunities for women. Dahlstrom's (1996) research in Troms County asked young men and women why they leave. Rural women have fewer opportunities than men to participate in all sectors of rural life and, consequently, moved for schooling or for service sector jobs more available in urban centres. Dahlstrom concludes that the "more traditional and less flexible" men (1996: 269) who remain behind are increasingly marginalized both in terms of economic opportunities and their chances to find wives, a point also made by Hanssen (2000:70). Finally, Solvang's (1997) examination of out-migration from her native Arnøya between 1980 and 1997 documents a 43.8 percent decline in the island's population (i.e., from 837 in 1980 to 470 in 1997) and suggests possibilities for future development, such as in tourism.

[3] The photo actually shows 71 students, 13 of whom came from the neighbouring islands of Laukøya and Vorterøya. Except where noted, my comments focus on the 58 Arnøy students.

numbers of school age children. One wonders how Lauksletta or Akkarvik might have fared if their school children had remained and raised families? Although the communities would have been larger, one must add that the existing, relatively homogeneous economy has fixed possibilities, which have long necessitated a certain level of out-migration.[4]

So where are the 58 students now? The largest group (24 of 58) are scattered throughout Norway. Several of these 24 live in the developing inner fjordal communities of North Norway, towns such as Nordreisa. The second largest two clusters live in the *fylke* capital of Tromsø (15 of 58) and in the kommune capital of Skjervøy (10 of 58), suggesting a desire to remain near Arnøya yet in a less isolated setting. Seven (7 of 58) live in Olso, and the same number (7 of 58) live on Arnøya. Now I could have reported that eight, not seven, lived on Arnøya in 1995, because that summer an Årviksand woman who had lived for a number of years in Oslo returned to her native Årviksand, with the intention to remaining permanently. Although a shortage of available housing forced the woman, her husband, and young daughter to live temporarily with her mother, both the woman and her husband considered Årviksand a better place to raise their daughter than the Oslo area. The woman's urbane husband (unable to work because of an occupational injury) appreciated Årviksand's untroubled society and exceptional natural beauty. Yet after a few weeks, the woman complained that, in her words, *it's too quiet here, I miss the city*. The couple returned permanently to Oslo in late summer, and I have, therefore, included her with the Oslo cluster. The final two small clusters of the 58 include one (1 of 58) who lives in Sweden, and five (5 of 58) that I could not locate.

The answer to my wife's question about where the Class of 1974-75 now live explains Arnøy's declining population. Simply put, of the 58 students, only 7 remain on the island. The rest have left the island, in a pattern of youth out-migration that threatens the long-term viability of the whole island, and most especially the immediate future of two of the island's four villages.

Where Did the Class of 1974-75 Go?

The answer to this question is not entirely clear from my data, although two of the larger clusters of out-migrants, people who emigrated to Tromsø and to Skjervøy, illustrate a pattern of chain migration, insofar as they followed others from Arnøy's

---

[4] Solvang (1997) reports varying levels of out-migration between the communities of Arnøya. Thus, between 1980-97, Akkarvik lost 64.7 percent, Lauksletta 51.6 percent, Årviksand 46.1 percent and Arnøyhamn 25.5 percent of their populations. These variations are not easily explained and Solvang attempts no explanation. I attribute Akkarvik's decline to the 'bust' associated with completion of the Trolltind telecommunications facility, the end of post-war reconstruction, and, more recently, to the failure of its fish plant. Much of Årviksand's out-migration was caused by the fishery crisis of 1989-90. The fact that Arnøyhamn lost the fewest people probably stems from its more generalized, service-based economy. Finally, the over 50 percent reduction of Lauksletta population is more difficult to explain, but is most likely linked to avalanche politics, as discussed in Chapter 6.

diaspora who moved to both places following the forced evacuation of the island by German troops in the autumn of 1944. Significantly, both Tromsø and Skjervøy, like the inner fjordal towns of Storslett and Nordreisa, mushroomed following the Second World War, and now offer more diverse economic and social opportunities. And as suggested above, the fact that many have moved to larger centres within Troms Fylke suggests either a desire to rejoin others from Arnøya who moved to places such as Skjervøy after World War Two, or simply the desire to live and find work in a larger place. But other than these generalizations, students from the Class of 1974-75 appear to have scattered throughout Norway.

A somewhat clearer pattern emerges when one asks what the students from the Class of 1974-75 are now doing? Most (16 of 58) work in various skilled trades, such as electricians or mechanics. Fifteen (15 of 58) work in the service sector, as teachers, nurses, or social workers. Twelve (12 of 58) work in businesses, ranging from accounting to tourism, and ten women work primarily as housewives. Only six (6 of 58) work in the fishery, either as fishers or plant workers. Five (5 of 58) work in assorted occupations and my informants were uncertain what six of the 58 did for work. But once again, only six work in the fishery and of people who returned to Arnøya (recall seven of 58), only two work in the fishery, one as a fisher, the other works in Arnøybruket (Årviksand's fish plant).[5]

Comparing the numbers of people who left Arnøya in the 'class of 1974-75' with the official statistics from the *kommune* (municipal) or *fylkes* (county) is difficult because the national data base does not break down statistics into micro regions such as Arnøya, that exist within *kommune* boundaries. Excellent as the data are from Statistics Norway (*Statistisk sentralbyrå*), the unit tallied is the *kommune*, and these numbers actually show that in the year 1974, the *kommune*'s population increased by 74 people. That is, in 1974, 233 people moved into the *kommune*, while 159 left it! However the same source records a net decrease of 31 people in 1975; that is, 139 people moved into the *kommune*, while 170 left, and presumably some of the latter number were from the 'Class of 1974-75.' Moreover, during nineteen of the twenty-five years between 1976 and 2001, Skjervøy *kommune* recorded a net decrease in in-migration, that is, fewer people moved to the *kommune* than left it. And finally, the same source records that in eighteen of the same twenty-five years (1976 to 2001), the *kommune* experienced a net decline in population growth, a figure based on the excess of births over deaths, and net migration. During the same twenty-five year time span, the *kommune* population decreased from 3677 in 1976 to 2996 in 2001, while during roughly the same period, the population of Arnøya plummeted from 916 in 1975 to 446 in 2001.

---

[5] The occupations of the 'Class of 1974-75' generally correspond with the responses from 42 children in my school questionnaire. Most of the children who claimed to know what they wanted to do as adults seek careers unavailable on Arnøya (ranging from airline pilots to horse riding instructor). Only three of the 42 children (7 %) want to be fishers, and 1 (2 %) to work in a fish plant. However, a surprising 5 youngsters (12 %) wanted to be farmers!

Why Did They Leave?

As I concluded in Chapter 5, labour shortages threaten today's fishery, and the history of this problem of recruitment dates from the 1970s. Why is it that beginning in the 1970s, youth turned their back on the fishery and many, by implication, on Arnøya?

There are several answers to this important question but the one most often heard is that elders (parents, teachers, and other adults) advised youth to seek careers outside the fishery. The youth listened obediently and left Arnøya. I heard this explanation time and again, but of course the explanation itself begs the question of 'why' such advice was commonly given at the time? Here, it seems apparent that in the modernizing sociocultural context of the 1970s, the advice of elders convinced youth that 'modern' jobs, rather than those in the fishery, were indeed preferable.[6] Moreover, at the time, people began to consider their lives, both in relation to familiar local realities and to the faraway (yet encroaching) world beyond Arnøya. This new reflexivity placed the once automatic entry into the fishery (usually on the day of a boy's confirmation into the Lutheran church) under a new and increasingly critical lens. Late-modernity meant reflexivity and choice, which gradually fuelled changes in local expectations of life and work.[7]

One man recalled his parents' efforts to discourage him from entering the fishery, as follows. *It's about 20 or 25 years since my parents told me don't ever be a fisherman, no good, and so on and so on. And they work hard to have me do something other than fishing. Me, aged 17 or 18, I wanted to be a fisherman. That should be my life but my parents say something else and I am not a fisherman. If my son asked me some day what do you think, should I be a fisherman? I will say, if you want to, go ahead. You just have to think over it before you start. It's hard work but there is good money and a good life.*

An Årviksand fisher recalled the kinds of conversations between local youth during the 1970s that both maligned fishing and fuelled out-migration. *They say there were not women in this place, that they go to school and that there was no work here then. They didn't want to be in the fishing boat. They said they got seasick. They didn't like fishing work, they now say it's okay. But they didn't then. They wouldn't come back now, as they have work elsewhere.*

---

[6] Writing about fishing communities in Finnmark, Hansen (1997) found that young people leave to obtain work not available in fishing villages. There are many similarities between Hansen's findings (despite different methods) and those reported here, although he makes no mention of the importance (if any) of return migration to Finnmark.

[7] According to Giddens (1991), choice and reflexivity are central characteristics of 'late-modernity,' contrasting with 'traditional' societies, where individual actions were not necessarily analyzed. Similarly, Bauman (1998) argues that the freedom to move is a central value in modern society. Related research by Wiborg (2003) interviewed 50 Bodø university (Norway) students from rural areas about their move and the meaning of home to them. Mobility and self-development were of primary importance, while their relationship to home was more 'complex, ambiguous, and dynamic' (165).

One man, who returned to Arnøya for holidays in the summer of 1995, recalled that in the 1970s, *I was glad that I was not going into the fishing boat. It was my opinion before I left the school in Arnøyhamn that I would not go into the fishery. And most of my friends have taken the same decision. We don't talk so much about it, we looked at our fathers and saw how they worked... very hard work... Hard work, it's a special work. It was the usual way to live here in 1950, 1960 and the years before that. Because their fathers [i.e., his grandfather] decide that's the only way to live here.* Thus, very typical of the 1970s was the advice parents gave children to seek occupations outside the fishery, as another man remembered: *Most parents told their kids to get away from Arnøya, get an education and a job. They didn't say anything. I got [had] 7 years in Lauksletta, then one year in Arnøyhamn, when I was 14, I went to Harstad for 2 years and then 2 years in Tromsø.*

This man's brother recalled the influential role played by school advisors, themselves often from southern Norway, who stressed the notion of choice, and of obtaining the *right* career when recommending careers other than fishing. *You see in the old days everything was focussed on the sea, to the fishery. We still look to the sea; that is normal [natural]. But when I was young, in the 1960s and 1970s, we had school advisors who would help the young get [i.e., select] jobs. And [they told us] you should see it in yourself what kind of job you were made for. There was a job for you. But if you go to a place in Norway called Sunndalsøra where we have one of the biggest aluminium factories, 90 per cent of the people work there. And whatever the advising people said, the young people got jobs in the aluminium. So these people in the 1960s should have thought about what they were saying. They should have said that if you want to stay in Lauksletta, 90 per cent of you are going to be fishermen. And that is natural. But they [advisors] say people are made to do different jobs and so, statistically, in Norway, only 20 per cent of people are fishers. I think these advisors led people away from the fishery, away from the normal. Clearly, the educational system, with its larger schools and school advisors encouraged youth to leave the fishery.* The same man recalled how youngsters of the time admired and may have consequently been influenced by the teachers who came to teach on Arnøya. In his words, *when I was in school, the teachers were from down south, from Oslo, Tromsø. We looked up to them and said 'okay, they are very good, they like football, and all the society around here did the same.*' I suspect that for Arnøy youth, these southern teachers were the physical embodiment of the new world just beginning to appear on local televisions, privileging the message of the teachers that preferable employment and lifestyles existed elsewhere. Yet it is also clear in the narratives just presented that to some extent it was the young as well as their parents who rejected the fishery, creating the 'chicken and egg' question of whether it was the advise of elders or the agency of youth which led young people to seek careers far from fishing and, consequently, from Arnøya.

In addition to the advice of elders and/or the agency of youth, schooling played an enormous role in discouraging the fishery. For many years, the educational system has required all young people to leave Arnøya after completing junior high school in Arnøyhamn. Education fuels out-migration, both by requiring youth to leave after junior high school and by training youth for careers which cannot be practised on Arnøya. In the words of the man quoted directly above, *another logical theory [to account for youth out-migration] is that in all the schools, you have to go away from this place to get an education. All are between 16 and 20 years. In that period you start to be an adult. You establish. If you're in Tromsø, you go to a pub; you do things that are urban. You get away from your roots. And that's happened with most of the young who get an education; they get a way of acting that is urban. That is normal.*

Similarly, a local woman who returned to her native Arnøya where she could no longer practice her occupation said, *if they [students] go to Tromsø, or other school places and left Årviksand permanently met somebody with a special education and wants a special job, they say, you can't go to Lauksletta with a special job. What did I do? I was a laboratory technician, I can't be that here, it would have to be Tromsø or somewhere like that. So, you see, I have been educated away from this kind of society.* An older Årviksand man, typical in so far as all his children followed the new roads (discussed in Chapter 6) away from Arnøya recalled, *the road [to Årviksand] came through in 1969 [others say 1968 or 1970]. All my children went 6 years to school here in Årviksand, then 3 years in Arnøyhamn, then høgskole [post secondary education]. Children had (by law) to attend* ***ungdomskole*** *[junior high school], which began in Arnøyhamn in 1974. Before that, children went directly into the [fishing] boat, but once* ***ungdomskole*** *came to Arnøyhamn, it was education, not fishing, in Arnøyhamn.*[8] Thus, in addition to the advice of elders, and to changing perceptions among youth, the educational system and the island's new road system encouraged youth to turn their backs on the fishery and on Arnøya.

Slowdowns in the fishery also influenced the decisions youth made, although it also appears that some youth may have left even during good fishing years. One man recalled, *they [youth] moved to go to school, and when the fishery was poor in the late 1970s and early 1980s, many sold their boats and moved away, to find an easier way to live. I remember about 1975, all the 1970s, it was very good fishing. Frode would remember. But the fish was all caught up. You see in the 1970s, it was free fishing [i.e., prior to the vessel quota system]; you could take as much as you could catch.*

---

[8] Unlike in North American education, where young people do not necessarily make career choices until the end of high school (i.e. at age 17 or 18), Norwegian youth make preliminary career choices at around age fourteen or fifteen, going on to videregående which may be either technical (i.e., agricultural, electrical, and others) or academic, the latter potentially leading to university. Thus, youth leaving the Arnøyhamn high school not only leave their island but also follow career pathways.

I would not want to create the impression that in the 1970s or since, adults callously banished youth from the island. As now, the adults of Arnoya wanted the very best for their youth, and seeing 14 or 15 year olds leave was devastating. One Årviksand woman remembered that *it was very difficult when the kids moved away. We considered moving to Tromsø to be close to the children.* Her sentiments echo many others I heard.

Contemporary Out-migration

Although more could be said about why the Class of 1974-75 turned its back on the fishery and left Arnøya, the problem of youth out-migration continues, albeit on a reduced scale. One reason why youth leave is that the Norwegian government provides generous financial support (i.e. *lånekassen*, or student loans)[9] enabling youth to pursue higher education. The lure of life beyond Arnøy's arcane beauty continues to pull youth from the island. One 19 year old man explained his plans to leave work at Arnøybruket (and Arnøya) in late summer of 1995: *I go in the army and after that I want to go to school to get a better job. I have not tried working on the fishing boat but know I get seasick. The money is good. I am 19 years old and most of my friends agree. A job on Arnøybruket over a long period of time is not good. It hurts the back. I want to work with my hands but not in a fishing boat or plant. I don't want to live in Årviksand, there's no future here. If you live here you must work in the fish plant. I want to live in a larger place like Skjervøy or Tromsø. Most of my friends agree. They go to school there, meet a girl and they don't come back.* The young man's explanation for his imminent departure is typical of many others that I heard, right down to the detail of seasickness. Youth who opt out of the fishery (even those working ashore in fish plants!) invariably use the excuse of seasickness. Yet virtually all-veteran fishers I interviewed recalled suffering from seasickness at the beginning of their careers. In one Årviksand fisher's words: *I know of only one or two old people [fishers] in Årvik who have never been seasick!*

My 19-year-old informant (just quoted), who planned to enter the army in the autumn of 1995, is also typical in his quest for work outside the fishery. His career plans resemble those heard by Gry Paulgård in her study of Honningsvåg (Finnmark) youth.[10] Paulgård found that Honningsvåg youth prefer 'modern' (albeit *seasonal* work) in tourism (such as at nearby Nordkapp hotels) to 'old-fashion' *permanent* positions in the fishery.

An astute Arnøyhamn man attributed the island's low unemployment rate to out-migration. In his words, *we have solved the unemployment problem by sending*

[9] In Nord Troms and Finnmark, students can write off 10% of their student loans up to a limit of 16,000 NOK per year.

[10] See Paulgård (1999) and Apostle et al (1998: 327). For reasons similar to those proposed here, Hanssen (2000) found that what I call global pulls are convincing young people (including fishers) from rural Finnmark to leave their communities.

the youth away. There are two main reasons why youth leave us. One is that work with fish today has a negative sound and the other thing is that they demand more of our [Arnøy] society than we can offer. Bigger places can offer more. But when they get families and get older, I hope that they will come back. And at that time we have something to offer them, employment and so on.

Although local opinions about the fishery have improved since the 1970s, youth continue to seek work outside the fishery.

Global Pulls

Anthropologist Ottar Brox devoted many years to the problem of explaining "the variation within the North Norwegian provinces, for example, why the population grew in one fishing village, but declined in another" (1996: 110-111). Brox explained differential out-migration by whether a person had a social or economic niche (that is, as wife, farmer, crewman, and so on) in his/her local community.[11] Although I think Brox's theory still explains some cases, most people I interviewed appeared to be less tied to the land and sea than they were during the 1960s, the time of Brox's research, reducing the value of such niches and consequently, of his explanation. Thus, by 1995, whether or not a person chose to occupy a local niche (i.e., a place in a fishing boat, farm land, and so on) was no longer, by itself, sufficient to deter out-migration, because the local niches Brox identified are no longer the coveted careers they once were. Moreover, youth face very real obstacles blocking careers in farming or fishing, such as, for example, gaining land or fishing quotas. But even so, my sense is that contemporary young people want

[11] Out-migration features prominently in Brox's theory, and is the focus of Pálsson (1974) application of Barth's generative model to Brox's work on migration and rural depopulation. According to Pálsson (1974), Brox's theory of out-migration includes many of the same variables considered by others but centres on how "individual choices reflect the structure of opportunities" (p. 30) which, in turn, affects his or her values and decision to remain in or move from one's local community. Thus, while many social scientists agree that migrants move to "improve their situation (mainly economically)," the decision to leave or stay hinges on whether an individual has a significant niche, or a place in their home community (Pálsson p. 25). Brox's research in his native North Norway discovered that persons owning (or eligible to inherit) land, possessing a berth on a fishing vessel, or married to persons who owned land, provided a niche that reduced the possibility that they would leave. For example, during Brox's brief field trip to the outside fishing community of 'Sand,' one of nine North Norwegian communities discussed in a 1972 paper, Brox observed that during the 1960s, 'Sand' was a viable community because it offered niches for its young men in the fishery. Pálsson concludes that for Brox, the "main factor behind depopulation trends in North Norway is the absence of niches for young unmarried people, since production is organized around the household unit demanding land, rights, capital and a certain amount of labour" (Pálsson p. 25). Thus, Brox's theory explains differences in community viability, assuming, of course, that such niches continue to be valued. More recently, Wiborg (2004) examines university student attachment to home 'places' and how such attachment affected the decisions of university students to leave (and/or return to) their home villages in northern Norway. The 50 students interviewed emphasized the symbolic and emotional attachments to home but tended less to see their home villages as a "material arena for social interaction" (p. 429).

something different than their parents or grandparents did. Today, youth seek livelihoods and lifestyles not obtainable on Arnøya. The reasons for what is essentially a cultural change in expectations are complex, suggesting a greater reflexivity among youth, and have been emerging since the 1960s. I maintain that by the 1990s, new global images increasingly challenged the meaning and value of local life. Put differently, the cumulative impact of a quarter century of cultural changes makes some locals, and especially young people, question the value of living and working locally.

As explained above, the Arnøya of Brox's time was road less and isolated. Læstadianism controlled various dimensions of social and religious life, causing people to look within themselves and their communities, rather than to the world beyond. However, the years between the late 1960s and my research in 1995 and 1997 was a period of immense change, affecting nearly every facet of local life.

Perhaps the most obvious symbol of change is television, which arrived on Arnøya around the time of Brox's visit. An older Akkarvik informant told me that in 1964 his father purchased the first television in the fjord, a 20-inch black and white set. Similarly, an Arnøyhamn man recalled that: *in 1966 or 1967, my uncle on the other side of the fjord got his first TV. It was a black and white, not so good picture but we were satisfied with it. It was closer to 1970, maybe 1968 or 1969, when we got our first television in this house.* Another Arnøyhamn man, quoted extensively in Chapter 2 in relation to Læstadianism, remembered that everything changed when his Læstadian grandfather died. As he put it: *in 1959, when grandfather died, we got a Christmas tree. And then we have radio. I remember very well the first television we had, that was in 1967 or 1968. And you have a very bad signal situation and so we run with a cable about 200 metres long and a big, huge antenna. Yah, we have a picture, we had a lot of pictures, on that screen. I think it was ten inches [others claim the first televisions were 19 inches], black and white.* Instantly, images of a world far beyond Arnøya began to appear in island living rooms.[12]

As a conduit of modernity, television encouraged reflexivity. One Arnøyhamn man explained: *we didn't know that in May they [other Norwegians] were sitting outside in Oslo and having beer. Suddenly we got TV and suddenly there is green grass, in Norway! They're talking about during the dark time in Norway people are getting [psychologically] sick, I don't know why. They cannot sleep or they sleep too much or they're getting wrong in the nerves. And in March and April they start*

[12] Relevant here is Selberg's (1993) paper on the introduction of television to Norway. Selberg argues that television ritualized everyday life, dividing each day between normal work time and 'sacred' or leisure time, commencing with the 7:30 PM news. Similarly, Davis (2003) describes her return to a rural Newfoundland community in 1989-90, and how villagers considered their village dull when compared with imported images from Detroit, viewed on their new cable television service. For a humorous autobiographical account of introduction of television to rural Tipperary Ireland in the 1960s, see Doyle (2005).

complaining about the snow and saying I don't think they'll be any summer this year. Why? They look on the TV and see that in the southern part of Norway, they have summer. Another, older Arnøyhamn man contrasted the pressures on contemporary Arnøy youth with the absence of such pressures fifty or more years ago, when young people entertained themselves and knew less about the outside world. Back then, he explained, *there was no one that has told you that there were "free time" problems. There was no pressure from the outside world about what you should do. It was natural to make your own social engagements. You didn't know about anything else. They could get together just to ski some distance to the youth house, have some entertainment. That was good enough. I have grown up with that environment. Never dissatisfied with it. Never known about the problem of free time.* Discussing the cultural influence of television with two men in Arnøyhamn, one observed that *television is a part of forming our lives. There was one lady here a few years ago who said, 'there has never been so much bad weather since we got television!' We enjoy it when they have some rain in the south!* To this, the other man added astutely, *when you see television images you may think that everywhere else things are better.*

In addition to changes which have made people reconsider their lives in relation to lives lived elsewhere, some locals blame television for changing the way people interact, eroding social capital, as one man explained: *before the television, people used to go visit one another, and talk about everything. And they take care of each other. They knew from day to day if a neighbour had a sick child, or if one of the sheep was going to have a lamb. They knew everything about each other. This could be both positive and negative. And now, we are sitting in each of our houses, looking at the television. And if we meet on the street we are not talking about each other but did you see that episode of something on last night. So people stopped being concerned about each other. If they can go in their own car, pick up some goods in the shop and then zoom back home, looking at the television.*

Similarly, an older Arnøyhamn man reminisced about how people once visited and gossiped aimlessly (in the Norsk vernacular, *slarve*, meaning benign gossip)[13] while, more recently, television has both reduced such inter-household visiting, and required that visitors have some purpose for their visit. In his example, *some years ago, two of my friends used to come here often. Recently, when they were visiting, I asked myself,* ***what*** *did they come for? It was more and more strange that people visited me just to visit. What purpose did you come for? It's more and more seldom that people go without a purpose [that is, people now go visit with a purpose]. [Formerly] you just had to talk. You didn't expect anything* ***(en forventet ingenting)****, any meal, or anything.* [14] This man's suspicion about *why*

[13] According to John Berger (1979:9) gossip such as this enables rural villagers to create a living portrait of their village.

[14] The transition this man describes recalls Ferninand Tönnies' (1957[1887]) distinction (often, I

Watching TV during confirmation party. Årviksand, 1995. Photo by Karen Olsson.

people now visit indicates some erosion of trust, and perhaps with it, of social capital. Television was introduced during a time of escalating individual autonomy, which together with other changes occurring during the 1960s and 1970s, allowed individuals to be less dependent on community.

By the late 1980s, broadcast of the national television (NRK, which now has two channels) network was augmented in many households by several channels received on home satellite dishes. As one Årviksand man said, *we get eight channels on television with [the satellite dish in] just the one direction but if it [the dish] was motorized, we'd get about 30-40 channels. Television satellite dishes came here about 6 years ago [ca 1989]. Johan had the first.* The result of this, in the words of another Årviksand man, is that *kids today are watching television, listening to the weather, and are thus very engaged in other things, than in social activities, much more of that today than 20 years ago. Everybody is watching television.*

Exactly one third of the 45 people in my adult questionnaire owned satellite dishes, and the number is growing. In my youth questionnaire, children reported that they watched as little as one-half hour of television daily, and as much as three

---

believe, misunderstood) between the Natural Will of *Gemeinschaft* and the Rational Will of *Gesellschaft*. Natural Will occurred when people assembled simply for emotional togetherness; whereas Rational Will pertained to association motivated by some specific, more tangible goal. The man just quoted appears to be making just such a distinction, between people visiting just to visit, and people visiting for some purpose.

to four hours, daily. The highest percentage (28 percent) watched between one to two hours daily, and mostly programs not originating in Norway.

Global images unintentionally challenge the assumptions of local life and identity, although not necessarily in ways directly related to out-migration. Moreover, imported television images (not simply advertising) link Arnøy people to the global culture of consumption. In the words of one man from Akkarvik, *television leads to pressure on parents to buy what others have. We get led by the television and should be able to judge for ourselves.* An Arnøyhamn man characterized the world of his eight-year-old daughter, claiming that *in her world, there is only Spice Girls and Barbies.*

Others trace more direct connections between imported images and local realities, including the fishery. An Arnøyhamn man attributed local difficulties recruiting fishers to negative national media images of the fishery. In his words, *maybe the media, television, maybe the people in Oslo, they say that it isn't good enough for younger people to be fishermen, lower status. It's not important enough.*

I contend that ever-increasing exposure to external influences whittles away at the very foundations of living and working locally, lessening the applicability of Brox's niche theory. Of course some youth remain, to live and to work, suggesting that the increasing impact of global images is powerful although, I emphasize, not yet all encompassing.

Of course the people of Arnøya, particularly young people, are not passive players in all these developments. Put differently, television or indeed other global pulls are not like some Pied Piper of Hamelin of the German Legend, leading dazed youth away from Arnøya. Instead, young people are well aware that the fishery is the foundation of their island, yet an arduous and economically risky way to make a living. And for most, to stay means to fish, or to work in a fish plant. According to Gry Paulgård (personal communication 1995), Honningsvåg *youth know their experiences; they know what they're talking about. It [the fishery] is cold, it's hard work, you're freezing. Some of their elders have occupational problems, bad backs.*

An Arnøyhamn man made the same point, although in the context of difficulties of fishery recruitment from the perspective of youth. He said: *to go in the fishing boats, it's very hard. You look at the sea; see the small boat, rocking up and down. And we have small boats like Alfredson, Arnøytind, and it's too small to get attractive work places. If you have boats that are trawlers, out at 7 o'clock, back at 7 o'clock, but in the small boats you have work all the time. They say that in the fishery it's hard, it's wet, and tastes bad.*

Moreover, the very real difficulties of recruiting young fishers are not related to money, as the salaries of most fishers compete with lucrative employment on land. But working at sea means one misses the growing attractions on land, even though, admittedly, not all these are yet available on Arnøya. The man just quoted explained: *young people are not going into the fishery because of money. If you're*

Confirmation cream cakes 1995. Photo by Karen Olsson.

on a boat, you can't be in the pub, you can't be in discotheques ***(diskoteket)****, you can't be in sports, you're away from all that.*

A defining character of our age is choice (see footnote 6, this chapter), or at least the possibility that one has several alternatives to any decision. Unlike their grandparents, who had few options to the *fiskarbonde* adaptation, the youth of today are surrounded by intrusive objects and ideas, and by the understanding that there are, and indeed should be, several alternatives to any decision. Thus, globalization exposes today's young people to images opposite to the fishery, and to living on Arnøya. Once again, Gry Paulgård found much the same in Honnigsvåg. She said: *When youth have the possibilities to choose, they don't choose fish. That's the bottom line. I also found (in Honningsvåg) lots of young people who waited for jobs at the hotel. There were jobs at the fishing plant near where they lived but they didn't want these. They just waited, laid in bed. Also, my Honningsvåg youth are very negative about everything that has to do with fish. They don't eat fish; they eat hamburgers and pizza.*

Since the 1970s then, there has been a steady drain of young people from Arnøya. At times this drain has been a trickle, at others, a deluge, such as with the Class of 1974-75. What appears certain is that constant exposure to a world beyond Arnøya fuels a desire to choose careers and lifestyles beyond Arnøya and that out-migration is motivated by more than just economic goals.

Reducing Out-migration

Dissuading youth from permanently leaving Arnøya is difficult, perhaps even impossible. Compared to the glitter and din of towns and cities, such as those portrayed on television or videos, the island's four small villages are dim and quiet. There are no shopping centres, sports stadiums, or cinemas on Arnøya. Arnøy people are acutely aware of the difficulties of keeping their youth at home and along with the uncertainties endemic to the fishery, consider out-migration a major threat to their island's future. Along with partners in government and industry, local communities are attempting to reduce out-migration through educational programs promoting the culture and history of rural areas such as Arnøya.

Coincidentally, one such educational program occurred during my fieldwork. The program, entitled *Kystsamfunn-Prosjektet* or the 'coast society' project, began in March 1994 with the aim of increasing the awareness of coastal culture and history among youth living in six *kommunes* in Troms county (similar programs were launched around the same time in Nordland and Finnmark counties).

Skjervøy was one of these six coastal Troms kommunes and the project was directed at remote coastal communities such as those of Arnøya. Several groups supported and funded the Kystsamfunn-Prosjektet, as the Skjervøy leader of the project explained. In his words, *the Kystsamfunn-Prosjektet was started by several groups. The fishery høgskole, the ministry of fishery (the Troms branch) which provides the way into the government (Fiskerisjefen i Troms), and then this Norwegian fishery union (Norges Råfisklag), and Troms Fylke, and the Institute for Fishery Research in Tromsø, and the Troms branch of the federal ministry of education. It is a difficult project because there are so many agencies involved. It has also been difficult to know which agency should fund what part.*

Next, the Skjervøy project leader explained the three main objectives of the Kystsamfunn-Prosjektet, beginning with the ultimate goal of reducing youth out-migration. In his words, *the aim is to increase the population in the coastal districts. That's the ultimate aim, to get a new start on coastal life. To keep young people home. There are three main aims. First, to strengthen the identity of people who live in these coastal areas, (done through the schools); second, to make youth aware of local possibilities in trade and industry, so that they're not only fishermen but also aware of new economic alternatives, and trade, marketing; and third, to better the recruitment to occupations based on natural coastal resources.*

During the project's three-year duration on Arnøya, the project leader in Skjervøy worked with schoolteachers and local communities to strengthen local awareness of identity and economic possibilities. Specific projects attempted to increase youth awareness about each community's particular economic assets and needs. Within Skjervøy kommune, for example, one project created grammar school texts emphasizing local culture and development possibilities; another collected local histories; and others toured fish plants, or took students out to sea on fishing boats.

The *Kystsamfunn-Prosjektet* was not the first program of its kind. In the early 1980s, a similar educational program began in the Lofoten islands and diffused to Arnøya. Older villagers visited the schools to teach youth about local practices, such as how to mend fishing nets or to prepare customary foods. Moreover, in 1996, over a year before the December 1997 end of *Kystsamfunn-Prosjektet*, discussions began for the Coastal Mobilization project, successor to *Kystsamfunn-Prosjektet*.

The existence of programs intended to enhance local identity and awareness, alongside continuing out-migration raises the questions about the effectiveness of these programs, or, put differently, the extent to which culture can be enhanced and preserved though educational programs.

Local opinions differ on this question. Perhaps not surprisingly, those directly involved with the project, were the most optimistic about its positive impacts. In 1995, the leader of the project in Skjervøy justified his involvement with the project and hopes for it, as follows. *I have chosen to do this because I think it will have an impact. But I am very realistic in a way because there are many things in the environment that says to me that we don't need more people in the fishing vessels. It's dependent on very many things that are national. The same in the traditional fish industry. You produce more and there are fewer employed. My aim is to focus on these possibilities that the resources give in all aspects of education. Why don't we have our children educated in marine biology and not only those from more central parts of the country? And thereafter we will have the possibilities to develop the resources and strengthen our local societies. And I want local youth from our area to have these jobs in the Spanish Embassy that tells about fish exports, because they know our area. And this will make it easier for the industry to change the direction. To have more of our coastal youth in jobs that are normally taken from the central part of Norway.* [John: that means that they leave their communities*?] Yes, but it would be worth it.*

Similarly, in 1997, near the end of the *Kystsamfunn-Prosjektet*, the director of the whole *Troms Kystsamfunn* project summarized the impact of the project as follows: *we know that in every county [in North Norway] the population has decreased. In the areas where we have worked, it's the same or increased. We are trying to change the attitude toward the older way of life. We are trying to make the government understand that we cannot live in these communities and produce as they do in the city. It has two big jobs. The first is to increase understanding.*

Among people on Arnøya, opinions differ about the efficacy of the Kystsamfunn-prosjektet. One young Årviksand woman supported the teaching of local history, claiming that it created *a feeling of belonging and pride* in place, a view similar to that of an Årviksand fisher. Teaching local history is important, he said, *because when old people die so dies knowledge and no one can continue without such knowledge.* One Akkarvik woman offered more qualified support for the project. *I think if you try to get the kids to choose educations that they can do [on Arnøya], it's possible to have an influence. But you can't force people to stay.* On the other hand, one man was much less certain about the efficacy of the project, as he explained in some detail: *you know, now you can sit in Arnøya watching CNN, live. You can switch and see stations from all over the world. Young people from 6 years until they leave school at 15, are sitting home watching TV. When I was 15 years of age, I had seen Tromsø, Skjervøy, and a little bit of the mainland, Nordreisa. I'd seen the fishing boats, the sheep, and my mother and father and people living around us. My son now is 15 years. I don't know his capability of speaking English but I know they start English in grade three. I started when we had two years left. Then we started learning English. So we are watching TV now and understand everything. I have to cheat sometimes and watch the text. But what I mean is that they knew so much more about the Norwegian society and the whole world society at 15 years of age, so much more than I knew. When two people are working in the* ***Kystsamfunn-prosjektet*** *and that they are putting it in the school and trying to let the young people know that maybe you can get a better life living and working in your home area. When they are putting it in the school, maybe you have one hour a week learning about cutting the wool off the sheep or maybe how to fix the fishing net, and so on. Okay, you get it like the rest of the schoolwork. You have one hour learning about your own place; your own culture and then you have the next hour learning English or mathematics. It becomes one hour of the schoolwork. Just something they are putting into the school day. They are taking it [local culture] out of reality. What they are learning about now was reality when I was young, and a generation before me. Reality now is CNN on the screen, email, and TV games. New cars. Their parents buy them cars when they're 18 years of age. [The* ***Kystsamfunn****] is talking about the past and as I see it, they're trying to get the young people in 1997 to rewind their minds and go back to the past in the 1960s. I don't think there's a chance.*

Although the goal of *Kystsamfunn*, summarized by the Skjervøy project leader as intending to make *youth proud of who they are and knowledgeable about coastal life,* is unquestionably beneficial, I am less convinced that programs of its kind actually reduce youth out-migration. Even so, I support such programs on the pedagogical assumption that rural youth may be more apt to master educational curricula containing familiar content, than those based on more imported concepts. Consequently, even though programs such as *Kystsamfunn* may not curb out-migration, they may help local youth learn based on familiar materials and, more certainly, will insure that those who leave will know where they came from. In the words of one Årviksand woman, the *Kystsamfunn-Prosjektet* sought *to give kids roots, not just wings!*

Thus, while there may be some educational value in programs such as *Kystsamfunn*, policy makers may want to consider whether funds allocated to programs aimed at reducing youth out-migration, might be better spent on improving rural infrastructure (roads, communications, schools, housing, and others) and on campaigns designed to attract some of those who leave to return home.[15]

It is important to recognize that beginning around thirty years ago, some of each generation who left Arnøya in their teens began returning permanently to the island in their twenties or thirties, to raise their families. Return migrants bring valuable urban skills and knowledge home to Arnøya and play important roles in the island's small villages. Most return primarily for two reasons: to avoid the high costs of urban living, and because they believe Arnøya is a much better place to raise children, than in larger centres. I recorded several narratives from such return migrants, of which the following two excerpts are representative. An Arnøyhamn man explained: *I came back here [Arnøyhamn] after I married. Me and my wife didn't like to live in the towns of southern Norway, we thought it best to live here. I had only intended to live in the south for one year. Many of us don't like to live in towns. I lived in Tromsø for many years. Both my wife and I didn't like it. My wife and I thought that if we had children, the problems about drugs and all that ... But I also said when I was young: I shall leave Arnøya. It wasn't place for a people. But now I know better. It's a good place to live. I was in the military; I was in Bergen, Stavanger. I was in Bergen in the Navy. I was travelling around Norway, in boats, diving, and I've seen a lot of things. I've seen how the other lives. And I think we are living as good as they are. It's the quality of life, life quality. I think we have it. We are a simple people, maybe. We are completely people of Norway. Yes, I think that.*

The second case, that of an Årviksand man, is complicated by the fact that he left the community and returned three times. Answering my question about his initial return, in 1982, he said, *I was thinking about fishing. Fishers have such free time on boat. I was 22 and had been away a year or so. I had been working slaughtering animals in Tromsø. I tried to see how to work and so on. I had been a fisherman before 1982, from 1979 to 1982. So I went back to the fishing boat. [He next left in 1983, to retrieve his wife]. I left in 1983 to get my wife. When I was in school (1984-5), I met her before I entered school. We came back here 1986. She was living in Trøndelag. I was gone (to Tromsø) between 1986-90, doing a variety of work, including drilling for water. I was in Tromsø, in 1988, I made a trawl net. There was more fish in 1988 than there was in 1987. I could see a future for fishermen. I called Einar [skipper of the Alfredson] from Trøndelag that was in 1990. I came back to work on the Alfredson. My work was a machine driver, heavy equipment. I didn't like it. It was lonely.*

Out-migration for varying amounts of time is now a 'normal' stage in the life cycle of many contemporary rural youth. Programs aiming to reduce it, while useful additions to local school curricula, fly in the face of strong contrary forces,

---

[15] Gmelch (1980) provides an overview of return migration. Other accounts of return migration to Newfoundland include Sinclair and Felt (1993) to Northern Peninsula, and Sinclair (2003) to Bonavista.

most recently global in magnitude. On the other hand, return migration strengthens Arnøya and boosts the chances that at least some of the island's four villages will survive. Consequently, the view taken here is that rather than attempting to curb youth out-migration, it may be better to improve rural infrastructure, thus building on Arnøy's already high quality of life, and perhaps encouraging some who leave for Tromsø and other centres, to eventually return.

CHAPTER 8

CONCLUSIONS

Two conclusions emerge from my examination of these voices from Arnøya. One has to do with community viability, the other with measures I believe need to be taken to address the consequences of out-migration. Both, but especially the second, may apply to other rural regions facing uncertain futures.

Before any discussion of community viability on Arnøya, it must be acknowledged that my comments *assume* that the resource base, in this case the Barents Sea Cod resource, persists, notwithstanding periodic crises.[1] Without fish, or some entirely new (and presently unknown) economic base, it is difficult to imagine how the communities of Arnøya could survive in the long term. All are aware of the enormous risks associated with the fishery, risks which are at once personal, collective, and environmental. Moreover, if Norway were to become more integrated with the EU through the EEA, and increase annual fish quotas to member states, this presumably would have consequences for the Arnøy fishery.

But putting aside this crucial and perhaps obvious qualification, the voices aired above suggest that there is no panacea ensuring community viability; many ingredients explain why one community survives and another withers. During the field research, the prospects for Lauklsetta and especially for Akkarvik appeared bleaker than those of the other two island communities, even though all populations on Arnøya have declined in recent decades, particularly since the fishery crisis described in Chapter 4. Akkarvik was unable to develop its economy beyond the flurry of activity associated with the post-war rebuilding of North Norway in the 1950s. Akkarvik's deep-water harbour certainly enabled the development of the local fishery, yet as we saw in Chapter 3, other ingredients necessary for such development failed to materialize. Thus, although the fishery might have provided an economic base beyond the post-war boom, a series of serendipitous events and decisions led the company of Leinebø and Larsen to favour Årviksand rather than Akkarvik. I need not repeat these coincidences of death and community loyalty but emphasize once again their importance in setting the stage for what followed: corporate decisions influenced *kommune* politicians who would increasingly favour Årviksand as the island's main fishing community.

---

[1] This assumption is, I realize, uncertain. Research by Myers (2003) concludes that industrial fisheries have depleted up to 90% of the world's predator species, including many of the species still fished in North Norway, and that we are, as Pauly et al (1998) conclude, 'fishing down the food web.' Consequently, it may be only a matter of time before the Barents Sea fishery follows others, such as Newfoundland and Labrador's Northern Cod fishery, as discussed in the Preface and Appendix 2, where cod have become commercially and perhaps even biologically extinct.

Lauksletta sibling group takes break from potato planting. 1995. Photo by the author.

One wonders whether Akkarvik's possible lack of social capital may have played a role in the community's wane, insofar as its post-war economic boom attracted people from various parts of north Norway who may not have had the time necessary to come to know and trust each other. Perhaps trust and reciprocity, the foundations of social capital, was less prevalent at Akkarvik, than elsewhere on the island.

In any event, as Akkarvik floundered, Årviksand held its own, galvanized by the decision of the company, and subsequently of the *kommune*, to christen it the fishing capital of the island. Of course Årviksand could not escape the dual fishing and banking crises of the late 1980s which would see the bullish times of the 1980s (*yappetid*) come crashing down, and would ultimately reduce Årviksand's fishing fleet from five relatively large fishing boats to one. Whether Årviksand's survival during these difficult years was aided by social capital is difficult to say, although in comparison with Akkarvik, Årviksand's population appears to have had longer roots on the island. If, and I stress if, social capital helped Årviksand survive beyond the crisis, its shallow, man-made harbour remains its on-going weakness, contrasting sharply with Akkarvik's deep fjord and suggesting again that single explanations fail to explain why certain places survive. Viewed differently, Årviksand's shallow harbour surrounded by its man-made breakwater, have become the *leit motif* for community survival, symbols of the very real assistance Årviksand continually needs from the state if it is to be the island's main fishing community. Those I spoke with (not all from Årviksand) emphasized that

Årviksand's small and shallow harbour, although too small to permit Russian trawlers to land their catches, has over recent decades figured prominently in regional fisheries politics, which in turn continues to attract media attention and state funding to the community. Viewed yet another way, as the island's main fishing community, Årviksand, like other single industry communities, is particularly vulnerable to periodic crises in the resource base.

Meanwhile, back at Akkarvik, closure of its school, continued deterioration of the vacant Leinebø and Larsen fish plant, and out-migration signalled a darker future. Understandably, Akkarvik people protested, attempting first to restore their right to land fish at the old Leinebø and Larsen plant, and later to resume ferry service to the community. Yet by then, their numbers were so small that kommune government allowed, however regrettably, the village to succumb by attrition.

The shallow waters of Arnøyhamn explain why fish continue only to be landed there and trucked to Skjervøy or Årviksand for processing.[2] Following the Second World War, Arnøyhamn's relatively centralized location, particularly after roads connected all communities, supported its position within successive *Kommune* General Plans as the service centre for the island. Arnøyhamn has the island's only church and graveyard, the only high school, an industrial zone, the clinic for travelling medical and dental personnel, a Senior Citizens apartment complex, a fox farm (until 1996), a sheep farm, and more; considerable diversity, at least by local standards.[3] It seems safe to conclude then that the key to Arnøyhamn's survival rests both in its relative economic diversity, and in *kommune* support for this.

The continuing decline of Lauksletta, perplexing in some ways, can be most likely attributed to two causes, as explained by the people I interviewed. First, although the community's fish plant is as old as Årviksand's, management at the Skaretfish plant was unpopular, explaining why most of the small and seasonal labour force commutes daily from Skjervøy. Consequently, the fish plant failed to attract local labour and become the economic foundation for Lauksletta that, to varying degrees, both Leinebø and Larsen and Arnøybruket have been for Årviksand. Moreover, filleting was not a prominent part of Skaretfisk's processing strategy, meaning that unlike at Årviksand, fewer women worked at the Lauksletta plant. Despite this, Lauksletta has a long history of entrepreneurship, with a popular retail shop, and successful salmon and sheep farms, each run by one of three young and energetic brothers, who along with a sister, live in the village, and with their families, are its nucleus. These siblings and their families form a close knit unit (a *klikk*, or clique); trust and business partnership overlays kinship.

---

[2] Following the field period described here (1995 and 1997), a man-made breakwater was constructed at Arnøyhamn, allowing larger fishing vessels to work from the community.

[3] Although Arnøyhamn *is* the most diversified community on the island, examples of economic diversity can also be found at Årviksand and Lauksletta.

Although the concept of social capital most certainly applies to this sibling group, examples of social capital are less apparent elsewhere in what is essentially an aging and fragmented village.

The second reason why Lauksletta's population continues to decline has to do with what I call (in Chapter 6) 'avalanche politics.' Prior to completion of the island's road system in 1970, travel between communities was by boat (or skis), meaning that winter avalanches off Singla Mountain were much less significant. Now, however, travel is by road. Although all island traffic must pass under Singla to take the ferry from the Skaret terminal, the degree to which Lauksletta is isolated by avalanche politics exceeds that of all other communities. As we saw, avalanche politics plus school centralization to Arnøyhamn led some young Lauksletta families to move to Arnøyhamn, and barring a tunnel linking the two communities, as discussed in Chapter 6, this trend will likely continue.

Although other examples might be mentioned, my analysis of the reasons behind differential community viability suggests that it is caused by an accumulation of various events and policies. A community's natural or environmental characteristics may be important, but need not be; the degree of social capital may be important but is difficult to accurately measure; and certainly *kommune* government policies and decisions, to some degree responding to political agitation from Arnøya, play a role in the fate of communities. Simply put, events accumulate, leading the places where people live in differing trajectories. Some of these events and decisions are local, the result of actions by individuals or groups, some far more distant, but with very real consequences radiating back to the island.

Even though I have restricted my comments above to the period following the Second World War, the time period emphasized in preceding chapters, it is appropriate to summarize the significant differences between the historic *fiskarbonde* economy discussed in Chapter 1, and the industrial fishery which emerged following the Second World War, described in Chapter 5. The decentralized, seasonally-ordered, and more generalized older fiskarbonde economy was probably more viable and unyielding to occasional resource fluctuations than is today's industrial fishery, insofar as crops and animals raised on the land might help families survive poor fisheries, and to some extent, vice versa. However, I would not wish to exaggerate the viability of the older pattern, and repeat that historic out-migration shows that there were real limits to the traditional economy. On the other hand, the viability of today's centralized and specialized industrial fishery, occurring eleven months of the year, depends on the health of fishery resources, even though the region is reinforced by the generous benefits of the oil-driven Norwegian economy. I have also questioned whether in the light of international comparisons, industrial fisheries can be viable, and concluded that they cannot. In short, comparisons between the historic economy and lifestyle and today's industrial fishery raise troubling concerns about the future.

Old barn with stockfish under netting. Lauksletta, 1995. Photo by the author.

A final comparison between the historic and present economies relates to out-migration, the topic of Chapter 7. It will be recalled that the impartible inheritance of the historic society meant that its reproduction required that children not inheriting the farm move elsewhere. In contrast, I have argued in Chapter 7 that during recent years inheritance and even finding a niche on the island are less relevant in decisions to leave: youth leave for school, for military service, or to see the world beyond their island homes. Consequently, as argued in Chapter 7, out-migration of youth is now a 'normal' part of the local life cycle of island youth. While economic and educational motives remain important in decisions to leave, the people I interviewed suggested that obtaining niches in local communities is less important now than it was a generation ago.[4] Today, youth want to leave island communities to see first hand the world of television and video imagery. Efforts to prevent this exodus, such as the Kystesamfunnproject, cannot be expected to stem the tide.

Rather than curbing out-migration, I have suggested that greater efforts might be made to convince some who leave to return.[5] The key policy implication here

[4] I consider this conclusion important and offer it as a friendly amendment to Brox's theory, as presented in Chapter 7, footnote 8.

[5] Obviously, I refer here to young persons or families, as opposed to elderly who may retire to the island.

lies in improving island infrastructure. A person in his/her twenties or thirties, who has lived for several years in Tromsø or Trondheim, and who decides to return home, will likely want schools, medical services, telecommunication technologies, roads, and other infrastructure roughly comparable to those he or she has come to appreciate in larger towns or cities. Improving rural infrastructure enables return migrants to enjoy the sociocultural rewards of coming home, without sacrificing the quality of services which they and their families enjoyed in urban centres. Return migrants bring precisely the kinds of capabilities necessary to diversify the island economy and society and are thus important to Arnøy's long-term prospects. Moreover and finally, if I am right that return migration promises to restore some of what has been lost through out-migration, those who remain on the island must be prepared to warmly welcome return migrants or others who chose to live and work on the spectacular island of Arnøya. This should not be difficult: speaking from my own experience, one could not expect a warmer welcome from people anywhere.

EPILOGUE 2003

In late September and early October 2003 I returned to Arnøya to go over my manuscript with those whose voices appear above. I first asked several Arnøy men and women if they would read the manuscript, and then sent it electronically to those agreeing to do so. Reading a long manuscript, in English, was I fully appreciate, no small undertaking, increasing my satisfaction when several local men and women agreed to do so. Some opted to read only particular chapters, others the entire manuscript. I stayed overnight with families in three of the four villages, and visited the fourth, Akkarvik. Peoples' daily work schedules meant that we often worked on the draft at night, or over the weekend. These sessions were enjoyable and extremely helpful to the overall project.

My sense is that most readers appreciated what I had attempted to do, and agreed with my argument about the importance of return migration to the island's future. Local readers also made a number of very helpful corrections, often on details of local history or Norsk spelling, all incorporated into this version.

The Arnøya of 2003 resembles that of 1995 and 1997, but there are changes, most illustrating various currents in the continued globalization of Norway's fishery. The island's two small fish plants, Arnøybruket at Årviksand and Skaretfisk at Lauksletta, continue to process fish. The survival of these two small plants is significant because several larger fish plants in Finnmark, such as the Aarsæther plant at Vardø, have fallen victim to globalization within the past few years. Moreover, Skjervøyfisk (the Skjervøy fish plant) and Refa ship repair (mentioned in chapter 5) were closed at the time of my visit, albeit temporarily.

Obvious yet surprising is that since the mid-1990s Skaretfisk is now the larger of the island's two fish plants, and the composition of its labour force possibly signals a different future for the local fishery. The success of Skaretfisk has been spearheaded by its young manager (whose optimism about the fish plant his father championed for so long explains his changing of his surname from Petersen to Skaret!), by the experimental collection system described in Chapter 5, and by a new reliance on migrant labour.[1] During my October 2003 visit to Skaretfisk, then processing sei, the plant employed approximately 55 people, an international labour force originating from fifteen different nations (Enoksen 2003). This globalization of labour, evident in Finnmark during the 1990s and just beginning on Arnøya during my fieldwork, is now necessary for the plant's operation. Immigrants from the Sudan and Ethiopia work more cheaply than locals, performing tasks many locals now prefer not to. In a little more than a generation, locals have abandoned the timeless fisher-farmer lifestyle of the past, and modern

[1] Despite this success, local fishers complain that Skaretfisk remains undercapitalised, and is thus slow to pay them for their catches.

work in the industrial fishery that replaced it; in rejecting both, people appear to prefer work in the service sector or business. Put differently, since the 1960s, locals have 'progressed' through three economic forms, each with its own relations of production and with the world beyond Arnøya.

When the recovery of Skaretfisk is viewed from another perspective, its reliance on imported labour forecasts a different relationship between the community and fishery. Although immigrant labourers depend on Lauksletta and Skjervøy for food, housing, and other essential services, the fish plant, rather than the community, is the focus of their life. It would appear then that the disembedded relationship between Skaretfisk and Lauksletta, obvious during the field period described above, is virtually complete. The existence of fish plants in communities such as Lauksletta, much like the 'home ports' of many contemporary freezer trawlers, becomes more incidental than explanatory, when assessing community futures.

Another kind of globalization of the fishery noted during my visit was the shipping of frozen, gutted fish to China, for filleting. Although the proportion of Norwegian fish exported for processing in China remains *small* (Torbjørn Trondsen, personal communication 2003), the potential for its increase could change *where* frozen white fish fillets are produced, reducing employment in coastal communities.[2] Although filleting has not in the past been important to Skaretfisk, in 2003 that company shipped for the first time 200 tons of its approximate 7000 tons of ground fish to China for filleting. Locals agreed that the quality of Chinese filleting was excellent.

In chapter 5, I expressed concerns about another form of globalization of the fishery, specifically, what at the time appeared to be a new reliance on Russian fish at Norwegian fish plants, particularly those in Finnmark. My concern here was that dependence on Russian fish could damage the vitality of Norway's coastal fleet but it now appears that my concerns were unwarranted: the Russians stopped delivering fish to Norwegian plants in the late 1990s. On the other hand, in early fall 2003, a Russian fish factory ship anchored at Akkarvik and purchased fish from locals, paying fishers 50 ore per kilo of fish more than did the local plants. Local fishers benefited from higher prices but local fish plants, most especially

[2] Although Professor Trondsen may be right, Canadian evidence documents the steady growth in the export of frozen, unprocessed fish to China for processing. It is my understanding that little, if any, filleting of groundfish still occurs in Nova Scotia, for example. Instead, fish caught are frozen at sea, exported raw to China, processed, and sold as boneless and skinless 'China cod.' One Canadian government website shows that Norwegian exports of raw cod to China increased from 8 million (US dollars) in 1998 to 63 million in 2002, and that Canadian figures for the same years rose from 25 to 79million dollars (see http://atn-riae.ca/asis/e3477.htm). A 2005 DFO discussion paper notes that globalization means "some Canadian companies are exporting their catch to China for processing. This means fewer jobs for domestic plant workers for processing this fish" (p. 13). Similarly, in 2005, the Clearwater Seafoods Company of Nova Scotia announced that is would close its (Arctic Surf clam) processing plant in Cape Breton and ship clams to China for processing (Bradley and Bethune 2005).

Arnøybruket, suffered from a loss of supply. By 2003, Arnøybruket's work force declined to around 15 local people. Indeed, when one compares the Arnøybruket of 2003 with that of 1995/97, one notes the loss of the filleting line and of the optimism of the mid-1990s. Perhaps anticipating this decline, the plant's former manager, a respected local man, accepted a job at Skaretfisk two years ago and has been replaced by a manager brought in by Norfra, Tromsø-based owners of Arnøybruket. Barring a local takeover of Arnøybruket, there are reasons to fear for the plant's future.[3]

Årviksand's artificial harbour and *molo* remain problematic, and Arnøybruket now relies primarily on the vessel Arnøytind, based in Arnøyhamn, for most of its fish. Meanwhile, the saga of the vessel Alfredson, chronicled in Chapter 5, continues. The vessel is now owned and skippered by Svein, the young engineer we met in Chapter 5, but the vessel now fishes without its coveted bottom trawl. The former skipper and co-owner of the Alfredson moved from Arnøya a few years ago while, on the other hand, three of the four crewmen of the fishing vessel Viggo Bakke (as described in Chapter 5) have moved back to the island. Out-migration appears to have stabilized and there have been several other cases of return migration.

Avalanche politics, discussed in Chapter 7, is an increasingly important issue. Given that six of the seven teachers at the Arnøyhamn School live in Lauksletta and commute to work daily, concerns about travel hinder schooling. During the 1999-2000 school year, 26 school days at Arnøyhamn were lost due to the threat of avalanches. Since then, an interim policy of 'winter schools' has emerged, whereby students who ordinarily travel to Arnøyhamn remain home and receive their school work by internet or by other electronic means. Of course tunnels promise a more permanent solution to avalanches. At a public meeting at Arnøyhamn in the spring of 2003, island residents gave top priority to a tunnel between Langfjord and Arnøyhamn, to benefit travel to Årviksand, and second ranking to a tunnel between Arnøyhamn and Lauksletta, the tunnel discussed in Chapter 7. The kommune government has accepted these local priorities and approved the projects in principle, even though their immense cost (and the need for tunnels elsewhere in Troms fylke) will likely require a wait of some years.[4]

---

[3] The comparative fortunes of Arnøybruket and Skaretfisk are evident in statistics available on-line (see www.ravninfo.com). Arnøybruket's labour force declined from 28 in 2000 to 14 in 2002, whilst Skaretfisk's numbers for the same years increased from 30 to 33. Similarly, Arnøybruket's total income declined steadily from 44,354,000 NOK in 1998 to 29,060,000 NOK in 2002, while Skaretfisk's total income steadily increased from 27,637,000 NOK in 1998 to an astonishing 77,421,000 NOK in 2002.

[4] More specifically, although both Arnøy tunnels (together called by the county the Nymoen-Årviksand tunnel) were estimated to cost 240 million NOK in 2000(Statens vegvesen 2000), and ranked third of eight tunnels considered most important in avalanche avoidance by the county, only 4 of the 240 million NOK was allocated for the project in 2004 (Troms Fylkesvegplan 2003).

The people of Arnøya are tenacious in their struggle to survive and well aware of the challenges that lie ahead. A final example from Akkarvik, a community I believed doomed, illustrates local determination and ingenuity. We will recall that the quay lay idle by the end of my fieldwork, with little sign that fish would ever be landed there again. However, in 1998 local fishers calling themselves the Akkarvik Fishery Service purchased the wharf and buildings from Dåfjord gruppen, and used volunteer labour to renovate the facility, enabling fish to be landed and trucked to Skaretfisk by 1999. These locals have also built a club or meeting place for fishers, suitably decorated with fishing motifs. Seven vessels (five from Langfjord; two from Akkarvik) fished from the harbour during the fall of 2003, and a dozen or so people were employed ashore, baiting long lines. Even though the question of whether Akkarvik or just its fishery will survive remains, the example shows a will to survive borne of necessity.

While the challenges posed by global forces remain very real and are escalating, my sense is that the people of Arnøya continue to believe that their small communities have a future; this, as much as anything, may help their island communities survive.

APPENDIX 1

METHODOLOGY

The idea for the Norwegian research arose after the Canadian moratorium closed the cod fishery in 1992, affecting much of Atlantic Canada, and most especially Newfoundland and Labrador. In early January 1995, my family (myself, my wife and two young sons) flew to Norway. The first month was spent in Bergen and Tromsø, reading about, and discussing coastal Norway with Norwegian academics. Social scientists at Tromsø, particularly Dr. Svein Jentoft, suggested that the island of Arnøya might be an appropriate location to examine the Norwegian fishery, the crisis, and perhaps, to learn lessons applicable to Atlantic Canada.

We moved to Arnøy in February, rented a vacant house in the island's largest village (Årviksand), and began learning about local life.

Understandably, Arnøy folk were initially reticent toward us and curious about our intentions. All members of my family helped in getting to know the community. For example, early in the fieldwork, the active tempo of work at the local fish plant led to a shortage of women available to work at the community's kindergarten (*barnehage*). This prompted local teachers to invite my wife to work for a few days at the kindergarten, introducing her to several local women, who in turn, introduced her to others. Our sons attended the kindergarten, and played daily with local children. Both learned to understand and speak Norsk, and their little playmates introduced us to their parents. Meanwhile, I talked with people at the fish plant, the small retail store, and in homes. Later, I accompanied some fishers to their nets, visited local farms, and so on.

Given my limited Norsk skills, I first concentrated on locals who were able to speak some English. Later, with the help of translators and with some improvement in my ability to at least understand some Norwegian, I was eventually able to interview and record a broad cross-section of the local population.

The Voices of Arnøya

The often heard (academic) concern about how anthropologists portray human communities, sometimes referred to as a 'crisis in representation,' requires brief comment insofar as my use of Arnøy voices to tell their story represents something of a departure, even from similar approaches of Lewis (1961), Crapanzano (1980; 1985) or Shostak (1981), three scholars who place local voices centre stage in their accounts. Goodall (2000: 140) defines voice as "the sound of the ethnographic world being *called* into being." Some anthropologists, such as Oscar Lewis, created named characters to represent various kinds of men and women found in particular cultural contexts.

However, this was not the approach taken here. Rather, I first present the context of whatever people will be discussing, and then use local voices to illustrate it. Arnøy men and women spoke to me on various topics, either in Norsk or in English. Following translation of the Norsk narratives into English, I made only minor editorial alterations, so as to present them as much as possible as I heard them. In order both to acknowledge yet mask the identity of some local translators, I have included the initials of those responsible for translating particular texts, usually written documents.

Finally, I include footnotes containing pertinent scholarly references or statistical documentation on or as near as possible to the relevant page in which it was mentioned. Such information is footnoted so as not to interfere with the centrepiece of the book, the voices of Arnøya.

The qualitative data or 'voices' I gathered amount to around seventy (70) hours of tape-recorded and translated interviews conducted in 1995, coupled with another twenty (20) hours in November-December 1997. Altogether, these recorded interviews are with seventy-one (71) people between the ages of 19 and 92, 50 are male, 21 female. As in most field data, the quality of the interviews vary; some informants limited themselves to the everyday topics, most were extremely perceptive and helpful. All people were cooperative and expressed varying degrees of interest in my work.

Because of our relatively short time Arnøya (six months in 1995; another month in 1997) and my limited Norwegian language skills, I decided to supplement standard participant-observation techniques with a questionnaire. The questionnaire was a revised version of one used for similar research in Newfoundland, as part of research on community 'sustainability.' I revised the Newfoundland questionnaire to suit the Norwegian context and my objectives, had it translated into Norwegian, and hired a local assistant to help administer it. Following a pilot and some revisions, I sought to have a minimum of ten per cent of adults aged eighteen or over complete it. Participation was voluntary and respondents were randomly selected from all four communities on the island. Each interview took an hour or more. We eventually completed questionnaires with over twelve (12) per cent of Arnøy adults. Significantly, given the initial reticence of some locals, the questionnaires helped introduce my project to islanders, creating the possibility for subsequent, more qualitative interviews. In the text, I have, for the most part, confined the statistical findings from the questionnaire to footnotes, so as not to compete with the qualitative narratives.

To better understand the continued emigration of people (mainly individual youth) from Arnøya, I also compiled a second, shorter questionnaire. My objective here was to learn about youth perceptions of local life, career choices, and the impact of imported cultural influences, such as television, on local youth. This youth questionnaire was administered in the island's two major schools to eighty-nine (89) percent of the forty-seven (47) island students between ages 7 and 15.

Finally, I returned briefly to Arnøya in the autumns of 2000 and 2003 to go over what I had written with locals. The 2003 visit was especially helpful because locals were able to review a completed manuscript. As noted in the Epilogue, virtually all of the suggestions I received from locals have been incorporated into the revised manuscript.

APPENDIX 2

THE COD CRISIS IN ATLANTIC CANADA

During the 1970s and 1980s, codfish became increasingly scarce in the once-rich waters off the Canadian province of Newfoundland and Labrador, and more generally throughout the waters of the Northwest Atlantic.[1] English and Irish settlers originally settled Newfoundland and Labrador to fish for cod, yet by the early 1990s, cod stocks plummeted so low that the Canadian government placed the fishery under moratorium in 1992. A brief account of what happened in Canada enables some comparison with the Norwegian case; both cases relate to similar processes now occurring among primary producers in other First World peripheries.

Although there are similarities between the Canadian and Norwegian fishery crises, there are also important differences of context, and these must be considered before discussing similarities between the two cases.

The severity of the fishery crisis in Newfoundland and Labrador, and indeed Atlantic Canada, varies according to when and which part of the region one is talking about. Generally speaking, however, the fishery crisis has been more severe in the more marginal northern area, along the coasts of Labrador and northeast Newfoundland, and less so further south, in the Maritime Provinces, where economic diversification cushioned some of the negative impacts. Also, because I am more familiar with the northern portion of the Atlantic region, that is, Newfoundland and Labrador, most of my examples are drawn from there.

As seen above, the cod crisis in northern Norway lasted only a few months. The situation in Atlantic Canada is more severe: there are actually fewer cod today in Newfoundland and Labrador, even without a cod fishery, than when the Moratorium began, on 2 July 1992.[2] Although both regions contain a remarkably large number of the same species (e.g., cod, herring, capelin, harp seals, and so on), the two regions have significantly different oceanographic and climatic conditions. The waters off northern Norway are warmed by the Gulf Stream, while the frigid Labrador Current chills waters off Newfoundland and

---

[1] Atlantic Canada includes the eastern Canadian provinces of Newfoundland and Labrador, Nova Scotia, Prince Edward Island, and New Brunswick. The three latter and older provinces are also referred to as the Maritime Provinces when distinguished from Newfoundland and Labrador.

[2] By 2005, thirteen years after the closing of the Atlantic cod fishery, numbers of cod were so low that the Canadian government considered designating cod a species at risk (DFO 2005), ultimately deciding not to do so for economic reasons.

Labrador. Consequently, ice bergs (calved annually off the Greenland ice cap) and Arctic (or Pack) ice produced annually in the Canadian Arctic, are carried south each year past Newfoundland and Labrador. Each winter local 'land-fast' ice freezes the sheltered bays and harbours of Northeast Newfoundland and the Labrador Coast to a depth of a meter or more. In sharp contrast, in northern Norway sea ice of any kind is rare or absent altogether (I saw none); and as we saw, cod fishing occurs during winter. Consequently, Newfoundland and Labrador's environment is more marginal than that of northern Norway, where cod stocks rebounded from the 1989 crisis within a few years.

The biological collapse of Newfoundland's northern cod fishery resulted from increasingly lethal fishing methods.[3] In the 1950s and 1960s, large factory freezer trawlers caught enormous catches of cod and other species. Canada's declaration of its 200-mile limit in 1977 excluded foreign trawlers, and this allowed the cod biomass to double between 1977-1985 (Hutchings 1999:262). Then Canada's offshore trawler fleet moved in, ultimately leading to the cod moratorium.

Although the Norwegian fishers I spoke with attributed the recovery of their fishery to effective fishery management by the Norwegian state, and contrasted effective Norwegian management with that in Canada (a position of some merit), I maintain that the failure of Canada's cod fishery is more complicated than management alone.

Even before Norway extended its offshore jurisdiction to 200 miles, in 1976, Norway signed bilateral agreements with the Soviet Union aimed at joint management and exploitation of Barents Sea fish stocks.[4] With the exception of several disputed 'grey zones,' Norwegian or Russian fishers fish Barents Sea cod exclusively, except when either country allocates small quotas to other nations. North Norwegian cod stocks 'hug' the coast while migrating between the Barents Sea and Lofoten Islands, while in the case of Newfoundland and Labrador, cod move well offshore in winter and then toward land in spring.[5] This leaves cod vulnerable to foreign fishing during winter, when they swim outside Canada's jurisdictional limit. This issue of 'straddling stocks' as it is called in Canada, combined with the continued fishery on the Grand Banks off Newfoundland by several European nations, hinders the recovery of fish stocks.

---

[3] This paragraph is based on Hutchings (1999).

[4] This section draws on Hoel (1991) and Apostle et al (1998).

[5] This sentence simplifies a very complicated reality: different interpretations exist regarding the extent of cod migrations in Newfoundland and Labrador, and about local and migratory cod stocks. For a thorough discussion of the history of local and scientific knowledge about the 'nature' of cod, see Hutchings et al (2002).

The difficulties of fisheries management in Newfoundland and Labrador are complicated by the province's late entry into the Canadian federation, in 1949. Newfoundland surrendered management of its ocean fisheries under term 3 of the 1948 Terms of Union. Ottawa would manage the fishery. There is now considerable agreement that Ottawa's management of the fishery has been disastrous, although there is less agreement about how to correct the situation. One correction would require Canada to extend its 200-mile limit in order to preserve straddling fish stocks but Canada appears hesitant to attempt such a bold, unilateral move.

The closure and/or reduction of the cod fishery throughout Atlantic Canada leaves a gaping void in the economy of many small communities, and initially placed eligible fishers and plant workers on income support programs provided by the Canadian Government.[6] What are some of the other consequences of the loss of cod in Canada? There are many so I limit this brief discussion to two, first, out-migration from Newfoundland and Labrador, and second, selected dimensions of fishing communities in the era after cod. This broad second focus covers issues ranging from the emotional impact of the cod crisis to disturbing new trends towards differentiation and privatisation in the new fisheries that have emerged in the wake of cod.

Out-Migration

A study by Dupuy et al (2000), conducted for the Government of Canada, concludes that while all regions of rural Canada are losing people, primarily young people, the Atlantic Provinces are losing more than other regions, with young Newfoundlanders moving to urban areas in other provinces. Similarly, a 2003 Royal Commission on Newfoundland notes that between 1991 and 2001, out-migration reduced the province's population by 10 percent (roughly 60,000 people), which the Commission concludes, is "unparalleled … for any country or province, except perhaps during wars." (Newfoundland and Labrador, 2003:36; 191).[7] Statistics Canada data shows that 47,100 people -- again, one out of ten people -- left Newfoundland and Labrador between 1996-2001 (Statistics Canada 2001). Most people leaving are young (between the ages of

[6] Canada's Department of Fisheries and Oceans (DFO) claimed in 2003 that between 1992-2001, Canada spent $ 3.9 billion on income support and economic development for the Atlantic fishing industry. Federal support was extended through several programs, beginning with the Northern Cod Adjustment and Recovery Program (NCARP), between 1992-94; the Atlantic Groundfish Adjustment Program (1993-94); the Atlantic Groundfish Strategy (TAGS) between 1994-1998; and Canadian Fisheries Adjustment and Restructuring Plan (1998-2001). Generally, these programs provided income support, development funding, and retraining programs to eligible fishers and plant workers involved in the ground fish industry and encouraged careers outside the fishery.

[7] Statistics Canada records that the population of Newfoundland and Labrador dropped from a high of 580,195 in 1993 to 512,900 in 2001.

15-29), and go west to the economically developed provinces of Ontario or Alberta, to find work. The Royal Commission just cited notes "the general tendency of many young people to leave, especially those with post-secondary education, combined with declining birth rates and the devastating effects of the collapse of the ground fish fisheries, with no recovery in sight, has depopulated many rural areas of almost their entire younger generations" (Ibid, p. 39).

Although people emigrated from Newfoundland and elsewhere in Atlantic Canada long before the fishery crisis, loss of the ground fish industry fuelled severe unemployment,[8] and out-migration. Levels of out-migration vary by region, as do levels of return migration. Perhaps relevant here is the conclusion of Dupuy and his colleagues that only 25 per cent of rural youth who leave return home, a percentage they consider inadequate to maintain the critical mass of rural populations (2000).

Research conducted in 1995 in one region, the fishery-dependent Bonavista peninsula of north-eastern Newfoundland, by sociologist Peter Sinclair and his colleagues (1998; 1999; 2002), also shows that it is the younger and relatively better educated persons who tend to leave their home communities, even when they were satisfied with living in the region. In an early discussion of this research, Sinclair et al (1998) report that although 30 per cent of Bonavista area residents planned to leave, those opting to remain were those most involved in the 'informal sector,' that is, those most embedded in inter-household networks or in the exploitation of local natural resources (p. 25). It is, Sinclair et al (1999) report, persons between 16-19 and with fewer commitments to the area who tend to leave, regardless of their involvement to the fishery (1999: 327). Most leave to find work, and more than half are willing to leave Newfoundland.

Another study, by Pinhorn (2002), examined aspirations of high school students in two rural schools in Newfoundland, finding that most students considered leaving their small communities a necessary future option, so as to insure a 'better' life, an opinion resembling the Norwegian voices aired in Chapter 7.

Fishing Communities Without Cod

As indicated above, most people have opted not to move, and instead to remain in a rural Newfoundland and Labrador that is less populated and more economically differentiated than was the case during the old days of the cod fishery. I return the issue of differentiation, below. First, however, as indicated above, the impact of Cod Moratoria on the individual and collected psyche has been great.

[8] Statistics Canada lists the unemployment rate in Newfoundland and Labrador as 21.8 in 2001.

Research I conducted in southeastern Labrador in the autumn of 2000 provides some examples of how people are coping without cod. Elsewhere (Kennedy, forthcoming), I present my results of this research at greater length; presenting here only a few examples. One woman I interviewed recalled her father's depression at the end of the cod fishery. She had never seen him cry until the moratorium; at its announcement, he cried "just like a baby." Another former Labrador fisher said that it took him two years before he could even *look* at the building where he stored his cod fishing gear. Like others, he is only beginning to deal with the loss of a way of life. The mayor of one Labrador community said that depression and low morale were common, especially during the early years of the fishery closure, between 1993-95. Idleness led to break-ins and brawls, particularly common among young men between 20 and 25 years of age.[9]

Perhaps because the cod fishery and distinctive seasonal moves to 'outside' fishing settlements persisted longer in southeastern Labrador than anywhere else in Atlantic Canada, Labrador people experienced a colossal change when the end of the cod fishery terminated their annual moves to 'outside' fishing settlements (Kennedy 1995). A way of life ended. Their depression resembled that which one feels on the death of a close loved one. Some of the scholarly writing on death provides an allegory for the end of the cod fishery. Thus, Boss' (1999) work on ambiguous loss -- where there is no closure on death -- may apply to the early years of the Moratorium, when all hoped the cod fishery would reopen. As the Moratorium wore on, certainty replaced ambiguity, with the denial, avoidance, and repression common to the grieving process (Spiegel 1977). It is only recently that people have entered the restitution phase, accepted the loss, and moved on with their lives (Nolen-Hoeksema and Larsen 1999).

Fortunately, some former fishers and cod fishing communities managed to cope with the end of cod. Research conducted in the Bonavista area of Newfoundland at the same time as Peter Sinclair's work, by Canning and Strong (2002: 330), shows that the collapse of the cod fishery has not produced "distress and despair" within families who remained in their home communities. Instead, the 'socially inclusive' character of Newfoundland's fishing communities explains the resilience of individuals and families who opted to not to move.

Some former fishers who remained in their home communities launched new careers outside the fishery, particularly in fields such as tourism. Still others continue to depend on the sea, fishing for shellfish or working in plants

[9] These findings resemble those of a study by the Newfoundland Division of the Canadian Mental Health Association (CMHA), which found depression, stress, boredom, family tension, and thoughts of suicide amongst former fishers in 23 fishing communities (1994).

processing shellfish. Overall, the fishery of Newfoundland and Labrador (and Atlantic Canada) has transformed from cod to shellfish.

Shellfish, specifically snow crab, Atlantic lobster, and shrimp, have replaced ground fish as the new focus of the fishery. The shellfish fishery employs fewer fishers than the historic ground fish fishery. In Atlantic Canada there were only 10,783 ground fish licence holders in 2000, compared with 17,200 in 1992, a decline of roughly 35 percent (DFO 2003).[10] Fewer fishers fish for fewer, yet more valuable fish.

The snow crab fishery occurs 200 or more kilometres offshore. Fishers use larger vessels than those formerly used in the cod fishery. Crab vessels are equipped with global positioning systems to determine the locations where iron crab pots are set and emptied, and to help navigate back home. And one reason snow crab stocks have increased is that Atlantic cod, which once ate young snow crab, are now rare.

Unlike the ground fish fishery, wherein value was measured by volume, value within the new shellfish fishery is based on the production of a high quality luxury food, exported to lucrative markets in the United States, Japan, and Europe. Thus, there is more wealth produced from the fishery today, than before the cod moratorium, even though there are fewer fishers, and increasing concentration of wealth in the hands of fewer people. According to the Federal Department of Fisheries and Oceans (DFO 2003), the Atlantic region now produces three-quarters of Canada's fisheries exports, valued in 2002 at $ 3.4 billion.

As noted, a major difference between the old cod fishery and contemporary crab fishery involves increasing differences of wealth and power. During the cod fishery, few fishers prospered; until the 1950s and 1960s, most fishers bartered their catches for goods needed to survive and most fisher families enjoyed an unpredictable and modest lifestyle. However, in the contemporary crab fishery, fishers who entered this new fishery in the 1970s, obtaining crab permits (and later, licences), are now wealthy. In the contemporary crab fishing communities of Atlantic Canada opulent homes and other displays of affluence now enable any visitor to know who owns a crab licence.

The lucrative nature of the new shellfish fisheries, coupled with cutbacks to federal programs aimed at assisting fishers in acquiring boats, invites various kinds of privatization, one example being 'trust agreements.'

Trust agreements are formal yet clandestine arrangements in which investors ('third parties, often fish processing or buying companies, but also funeral homes, and other investors) finance vessels or licenses, which are then fished by fishers, who effectively relinquish control of the fishery. Many fear

[10] Facts presented in this and the following paragraph appear in a Canadian Fisheries and Oceans summary available at www.dfo-mpo.gc.ca/media/backgrou/2003/cod-1_e.htm

that trust agreements could permanently separate licenses from communities, threatening community viability.[11] In short, since the expansion of the lucrative shellfish fishery, there has been increasing corporate interest in controlling licenses held by inshore shellfish fishers. Most fishers in the region oppose trust agreements. For its part, Canada's Department of Fisheries and Oceans remained silent on the issue until 2003, when the department agreed that such agreements ran counter to public policy (DFO, as cited in Allain 2004:45). Yet DFO has been slow to ban such agreements and as matters stand, trust agreements continue to represent a threat to the independence of rural fishers and to the viability of their home ports.

Although comparisons between Newfoundland and Labrador, and northern Norway are always subject to the particulars of history and context, the general contours of Arnøy case presented above relate to those occurring in other rural regions, such as Atlantic Canada. Of course the details of such comparisons are always unique. Thus, for example, although I found nothing like trust agreements in the Norwegian fishery, there the threats to community viability include issues such as fishing vessels or fishing quotas leaving their 'home' ports. And as noted, both in Atlantic Canada and North Norway, the exodus of young people from their home communities remains a concern, as discussed in Chapter 7. In short, the case of Arnøya, although unique in many ways, relates to others in which rural peoples face challenges to the viability of their way of life.

---

[11] This section draws heavily on the work of Marc Allain, Communications Director of the Canadian Council of Professional Fish Harvesters, the national body concerned with policies concerning fish harvesting. See Allain (2004; 2005); FFAW (2004); Fishingnet.ca (2004).

REFERENCES

Aas, S., 'North Norway - The frontier of the North?' *Acta Borealia* (1998) 1: 27-41

Adams, J. (Ed), *Fighting for the Farm: Rural America Transformed,* Philadelphia: University of Pennsylvania Press (2002)

Allain, M., Inshore fisheries: A Case to Follow, SAMUDRA, Report No. 39 (2004)

Allain, M., Sustainable Fisheries Policies, Policy Backgrounders, Prepared for Rural Communities Impacting Policy (RCIP) Rural Policy Forum (2005)

Amid, V. and N. Rapport, *The Trouble with Community*, London: Pluto Press (2002)

Andersen, R.R., Hunt and Deceive: Information management in Newfoundland Deep-Sea Trawler Fishing. In R.R. Andersen and C. Wadel (Eds), *North Atlantic Fishermen: Anthropological Essays on Modern Fishing*, St. John's: ISER, Newfoundland Social and Economic Papers, No. 5 (1972) 120-40

Apostle, R., G. Barrett, P. Holm, S. Jentoft, L. Mazany, B. McCay, and K. Mikalsen. *Community, State, and Market on the North Atlantic Rim,* Toronto: University of Toronto Press (1998)

Arbo, P. and B. Hersoug, The globalization of the fishing industry and the case of Finnmark, *Marine Policy* 21(2) (1997) 121-41

Arensberg, C., *The Irish Countryman*, New York: Natural History Press (1968)

Aubert, V., H.D. Bratrein, O. Irgens-Jensen, F. Fjellberg, and P. Mathiesen, Isolation and Integration: A Community Study in Northern Norway, Oslo: Institute for Social Research, Dept. of Sociology (typescript) (1970)

Austrem, L.M. and G. Ingebrigtsen, *Feskarbonden og andre Kvinnfolk*, Stamsund: Orkana Forlag (1992)

Barth, F., *Models of social organization*, London: Royal Anthropological Institute Occasional Paper No. 23 (1966)

Barth, F., 'Introduction' to F. Barth (Ed), *Ethnic groups and boundaries*, Boston: Little, Brown and Co (1969) 9-38

Bauman, Z., *Globalization: The human consequences*, Cambridge: Polity Press (1998)

Bentsen, C.J., 'Skromt-intematet på Haugues', *Årbok Menneske og Miljø I Nord-Troms*, (1985) 72-74

Berg, P.O., A. Linde-Laursen, and O. Löfgren (Eds), *Invoking a transnational metropolis: The making of the Øresund region*, Lund: Studentlitteratur (2000)

Berger, J., *Pig Earth*, London: Writers and Readers (1979)

Bjørklund, I., 'Fields of Interaction and Implication systems: Læstadianism and Social Change in the North of Norway', *Ethnos* 1-2: (1988) 63-77

Bjørklund, I., 'The anatomy of a millennarian movement. Some organizational conditions for the Sami revolt in Guovdaageaidnu in 1852', *Acta Borealia* 2: (1992) 37-46

Bjørklund, I., *Sápmi - Becoming a Nation*, Tromsø: Tromsø Museum (2000)

Blekesaune, A., 'Changes in Ways of Making a Living among Norwegian Farmers, 1775-1990', *Sociologia Ruralis* 31(1) (1991) 48-57

Blekesaune, A., W.G. Haney, and W.S. Haugen, 'On the question the Feminization of Production of Part-time Farms: Evidence from Norway', *Rural Sociology* 58(1)(1993) 111-29

Boss, P., *Ambiguous Loss: Learning to Live with Unresolved Grief*, London: Harvard University Press (1999)

Bourdieu, P., In J.G. Richardson (Ed), 'The forms of capital', In *Handbook of Theory and Research for the Sociology of Education*, New York: Greenwood (1985) 241-58

Bradley, S. and J. Bethune, 'Clearwater cutting jobs at C.B. plant', *Halifax Chonicle*, October 15, (2005) F1.

Bratrein, H.D., 'Capital influx and credit problems in North Norway through history', *Acta Borealia* 1: (1992) 39-45

Bratrein, H.D., *Karlsøy og Helgøy Bygdebok, Fra År 1760 til 1860*, Bind II, Hansnes: Karlsøy Kommune (1990)

Brooke, A., *Travels through Sweden, Norway, and Finmark, to the North Cape, in the Summer of 1820,* London: Rodwell and Martin (1823)

Brox, O., 'Social stratification in a North Norwegian Fishing community', In H. Hvarfner (Ed.), *Hunting and fishing*, Luleå, Sweden: Norrbotten's Museum (1965) 83-90

Brox, O., *Avfolking og lokalsamfunnsutvikling*, Bergen: Norges Landbrukshogskoke. Institutt for driftslære of landbruksokonomi, Memorandum Nr. 60, Avdeling for bygdesosiologi, Shrift Nr. 25 (1972)

Brox, O., *Nord-Norge Fra Allmennig til Koloni*, Tromsø: Universitetsforlaget (1984)

Brox, O., 'My life as an anthropologist', *Ethnos* 61(1-2): (1996) 105-20

Brox, O., 'The politics of common resource management', In Jentoft, S. (Ed), *Commons in a Cold Climate: Coastal fisheries and reindeer pastoralism in North Norway: The co-management approach,* Paris and New York: UNESCO and Parthenon Publishing Group (1998) 335-45

Bryon, R., 'Introduction', In R. Bryon (Ed), *Retrenchment and Regeneration in Rural Newfoundland*, Toronto: University of Toronto Press (2003) 3-13

Cadigan, S., 'Whose fish? Science, ecosystems ethics in fisheries management literature since 1992', *Acadiensis* 31(1): (2001) 171-95

Canadian Mental Health Association, 'Working it out, the challenge of change from within: A needs assessment for community self-help', Produced by Extension Community Development Co-operative, CMHA, Newfoundland Division, Health Canada (1994)

Canning, P. and S. Strong, 'Children and families adjusting to the Cod Moratorium', In R.E.Ommer (Ed), *The Resilient Outport*, St. John's: ISER, Social and Economic Papers No. 25 (2002) 319-41

Coleman, J.S., 'Social capital in the creation of human capital', *American Journal of Sociology*, 94: (1988) 95-121

Crapanzano, V., *Tuhami: Portrait of a Moroccan*, Chicago: University of Chicago Press (1980)

Crapanzano, V., *Waiting: The Whites of South Africa,* New York: Random House (1985)

Curtin, C., 'Back to the future? Communities and Rural Poverty', In C. Curtin, Haase, T., and H. Tovey (Eds), *Poverty in Rural Ireland: A Political Economy Perspective*, Dublin: Oak Tree Press (1997) 248-274

Dahlstrom, M., 'Young Women in a Male Periphery - Experiences from the Scandinavian North', *Journal of Rural Studies* 12(3): (1996) 259-271

Danielsen, R., *Norway: A history from the Vikings to our own times*, Oslo: Scandinavian University Press (1995)

Davis, D.L., 'In the beginning: Region, crisis, and occupational choice among Newfoundland's youth', In R. Bryon (Ed), *Retrenchment and Regeneration in Rural Newfoundland*, Toronto: University of Toronto Press (2003) 177-98

DFO, 'Current State of the Atlantic Fishery. Backgrounder', www.dfo-mpo.gc.ca/media/backgrou/2003/cod-1_e (2003)

DFO, 'Socioeconomic considerations to inform a decision whether or not to list 3 populations of Atlantic cod under SARA: Discussion Document', www.dfo-mpo.gc.ca/species-especes/cod/cod_morue_.pdf (2005)

Doyle, J., *A Great Feast of Light*, Toronto: Doubleday Canada (2005)

Dupuy, R., F. Mayer, and R. Morissette, 'Rural Youth: Stayers, leavers, and return migrants', Ottawa: Rural Secretariat, Public Works and Government Services Canada (2000)

Durrenberger, E. P. and G. Pálsson, 'Systems of production and social discourse: The skipper effect revisited', *American Anthropologist* 92(1): (1996) 130-141

Dyrvik, S., 'Farmers at Sea: A study of Fishermen in North Norway, 1801-1920', *Journal of Family History* 18(4): (1993) 341-56

Economist, 'Norwegian banking: Up the Fjord', *The Economist* Vol. 318(7689): (1991) 73

Eikeland, S., *Robuste Fiskevær? Selinvasjoneses Virkninger for Finnmarkskysten*. Oslo: Norges forskningråd (1993)

Elstad, Å., 'Changing labour roles in Norwegian fisherman-farmer households', In R.R. Balsvik and S. Gerrard (Eds), *Global coasts: Life changes, gender changes*, Tromsø: Kvinnforsk (1999) 19-29

Enoksen, R., 'Verden på Skaret', *Nordlys* Sept. 27 (2003) 40-42

FFAW/CAW, 'Inshore council endorses fisheries policy proposals' Dec. 2 (2004)

Finstad, B-P., 'Hanna Anny from the Ukraine - A pioneer in the fishing industry of northern Norway', In S. Gerrard and R.R. Balsvik (Eds), *Global Coasts: Life changes, gender challenges*, Tromsø: University of Tromsø (1999) 30-38

Fjørtoft, K., *Dramaet på Arnøy*, Oslo: Gyldendal (1981)

Friis, J.A., *Ethnographisk Kart over Finmarken No. 1*. Præstegjældene J.A. Friis, Cand. Theol. Udgivet af Videnskabsselskabet i Christiania, med Bidrag af Oplysningsvæsenents, Fond (1861)

Friis, J.A., *Ethnografisk Kart over Tromsø, Ant No. 2*. Præstegjældene J.A. Friis, Prof. Ved Universitetet I Christiania, Udgivet Paa Offentlig Betostning (1890)

Frimannslund-Holmsen, R., 'Comparative studies of peasant communities in Norway. A Personal Account', *Norsk Geografisk Tidsskrift* 53(2-3): (1999) 161-65

Frykman, J. and O. Löfgren, Culture *builders: A historical anthropology of middle-class life*. New Brunswick, N.J.: Rudgers University Press (1987)

Fugelsøy, M., *Skjervøy: Et Prestegjeld og et herred i Nord-Troms* Bind II. Skjervøy: Utgitt av Skjervøy Kommune (1974)

Gaffin, D., *In Place: Spatial and Social Order in a Faeroe Islands Community*, Prospect Heights, Illinois: Waveland Press (1996)

Gerrard, S., 'When women take the lead: Changing conditions for women's activities, roles and knowledge in North Norwegian fishing communities', *Social Science Information* 34(4): (1995) 593-631

Giddens, A., *The Consequences of Modernity*, Oxford: Polity Press (1990)

Giddens, A., *Modernity and Self-identity. Self and Society in the Late Modern Age*, Barcelona: Polity Press (1991)

Gmelch, G., 'Return Migration', *Annual Review of Anthropology* 9: (1980) 135-59

Goodall, H.L., *Writing the New Ethnography*, NY: Altamira Press (2000)

Gorter, W.T., 'Ethnic identity, resource management and nation building in Northern Norway', In J. Boissevain and J. Friedl (Eds), *Beyond the community: Social process in Europe*, The Hague: Dept. of Educational Science of the Netherlands (1975) 147-64

Granovetter, M., 'Economic action and social structure: The problem of embeddedness', *American Journal of Sociology* 91: (1985) 481-510

Gray, I. and G. Lawrence, *A future for Regional Australia Escaping Global Misfortune*. Cambridge: Cambridge University Press (2001)

Gullestad, M., *Kitchen-table society: A case study of the family life and friendships of young working-class mothers in urban Norway*, Oslo: Universitetsforlaget (1984)

Guttormsen, H., 'De Eldste Reinbeitedistriktene i Nord-Troms', *Menneske og Miljø - Årbok for Nord-Troms* (1993a) 14-17

Guttormsen, H., 'Nytt fra Skjervøy', *Menneske og Miljø – Årbok for Nord-Troms* (1993b) 91-93

Guttormsen, H., *Tidsvandring gjennom et kystculturlandskap* (n.d.)

Hache, J-D., 'Towards a Political Approach to the Island Question', In G. Baldacchino and R. Greenwood (Eds), *Competing Strategies of Socio-Economic Development for Small Islands*, Charlottetown, PEI: Institute of Island Studies (1998) 31-68

Hamilton, L.C. and O. Otterstad , 'Sex Ratio and Community Size: Notes from the North Atlantic', *Population and Environment* (1988) 20(1) 11-22

Hansen, J.C., 'Why do young people leave fishing communities in coastal Finnmark, North Norway?', In R. Bryon and J. Hutson (Eds), *Local Enterprise on the North Atlantic Margin*, Aldershot: Ashgate (1997) 225-51

Hanssen, K., 'Fisheries dependent communities in Finnmark: Dependence on fish or people?' In D. Symes (Ed), *Fisheries Dependent Regions*, Oxford: Blackwell (2000) 65-74

Haugen, M.S., 'Female Farmers in Norweign Agriculture', *Sociologia Ruralis*, 30(2): (1990) 197-209

Helland, A., 'Norges Land og Folk: Topografisk-Statistisk Beskrevet XIX Tromsø Amt', Kristiania: H. Aschehoug & Co (1899)

Henriksen, H., 'Drift igjen på Arnøya', *Nordlys* 28 December (1990)

Henriksen, H., 'Arnøya frykter nedbygging', *Nordlys* 7 Sept. (1991a)

Henriksen, H., 'Opprør på Arnøya', *Nordlys* 7 Sept. (1991b)

Herzfeld, M., *Cultural Intimacy Social Poetics in the Nation-State*, New York: Routledge (1997)

Hetland, P., 'Å bli fisker - formell og unformell sosialisering i fisket', In S. Jentoft and C. Wadel (Eds), *I Samme Båt: Sysselsettingssystemer i fiskerinæringen.* Oslo: Universitetsforlaget (1984) 36-46

Hetland, P., 'Pluriactivity as a strategy for employment in rural Norway', *Sociologia Ruralis* 26 (3-4): (1986) 385-95

Historisk Institutt, Folketellinga 1801- Skjverøy, Universitet I Bergen, Bergen 5014, (1801)

Hoel, A.H., 'Managing the Barents Sea Fisheries: Impacts at National and International Level', Paper presented to MAB Conference, Alta, Norway, 5/10/91 (1991)

Holberg, E., *Kulturminer på Arnøy og Laukøy i Nord-Troms*, Sørkjosen: Nord-Troms Museum II (2002)

Holm, P., B. Hersoug, and S.E. Rånes, 'Revisiting Lofoten: Co-managing fish stocks or fishing space', *Human Organization* 59(3): (2000) 353-64

Homans, G.C., *English Villagers of the Thirteenth Century*, New York: WW Norton (1975[1941])

Hovgaard, G., 'Coping with Exclusion - A North Atlantic Perspective', In N. Aarsæther and J. Bærenholdt (Eds), *Coping Strategies in the North*, Copenhagen: Nordic Council of Ministers (1998) 45-70

Hovland, E., H.W. Nordvik and T. Stein, Proto-industrialization in Norway, 1750-1850: Fact or Fiction?, *Scandinavian Economic History Review* 30(1): (1982) 45-56

Hutchings, J.A., 'The biological collapse of Newfoundland's Northern Cod', In Ommer, R.E. and D. Newell (Eds), *Fishing places, fishing people: Traditions and issues in Canadian small-scale fisheries*, Toronto: University of Toronto Press (1999) 260-75

Hutchings, J.A., B. Neis, and P. Ripley, 'The 'Nature' of Cod, Gadus morhua', In R.E. Ommer (Ed), *The Resilient Outport*, St. John's: ISER Social and Economic Papers No. 25 (2002) 140-85

Jakobsen, S-E., 'Political regulation and market allocation: Development trends in the Norwegian fish farming industry', *Norsk Geografisk Tidsskrift* 50(3-4): (1996) 187-99

Jentoft, S. and C. Wadel (Eds), *I Samme Båt: Sysselsettingssystemer i fiskerinæringen*, Oslo: Universitetsforlaget (1984)

Jentoft, S. and T. Kristoffersen, 'Fishermen's Co-Management: The Case of the Lofoten Fishery', *Human Organization* 48(4): (1989) 355-64

Jentoft, S., 'Hvor sårbare er fiskerimiljøene?' In S. Jentoft and C. Wadel (Eds), *I Samme Båt: Sysselsettingssystemer i fiskerinæringen.* Oslo: Universitetsforlaget (1984) 149-77

Jentoft, S., *Dangling Lines: The Fisheries Crisis and the Future of Coastal Communities, the Norwegian Experience*, St. John's: ISER, Social and Economic Studies No. 50 (1993)

Johnsen, H.G., *Kulturminnene på Skattøra: Ingen monumenter uten historie*, Tromsø: Troms fylkeskommune (1996)

Jørgensen, J.E., P. Jørgensen, T. Ottesen, and J.P. Jørgensen, 'Prosjektoppgave i Norsk med vekt på lokal samfunnshistorie og næringsutvikling i Skjervøy Kommune', (Faglærer: Helge Guttormsen) typescript (1995)

Kennedy, J.C., *Anthropological history and the fate of communities in the unknown Labrador*, Toronto: University of Toronto Press (1995)

Kennedy, J.C., 'Disempowerment: The Cod Moratorium, Fisheries Restructuring, and the Decline of Power among Labrador Fishers', In P. R. Sinclair and R.E.Ommer (Eds), *Power and Restructuring: Canada's Coastal Society and Environment*. St. John's: ISER (forthcoming)

Kvaal, G.A., *Karen Anna and her Siida*, Oslo: Pax Forlag (2001)

Lelé, S.M., 'Sustainable Development: A Critical Review', *World Development* 19: (1991) 607-23

Lewis, O., Children *of Sánchez: the Autobiography of a Mexican Family*, New York: Random House (1961)

Lin, N., *Social Capital: A Theory of Social Structure and Action,* Cambridge: Cambridge University Press (2001)

Lindkvist, K. B., 'Local Industrial Development in Finnmark and Company Ownership', In Aarsæther, N. and J. O. Bærenholdt (Eds), *Transforming the Local* Copenhagen: Nordic Council of Ministers (2001) 119-145

Mathiesen, P., 'Ethnicity as Pattern: Past and Present', *Acta Borealia* 1: (1990) 5-13

Mathiesen, P., 'Reconstructions of Relationships and Ethnic Boundaries', *Acta Borealia* 1: (1993) 37-44

Maurstad, A. and J. H. Sundet, 'The Invisible cod – Fishermen's and Scientists' knowledge', In Jentoft, S. (Ed), *Commons in a Cold Climate: Coastal fisheries and reindeer pastoralism in North Norway: The co-management approach*. Paris and New York: UNESCO and Parthenon Publishing Group (1998) 167-84

Mewett, P.G., 'Setting and Situation in Migration from the Isle of Lewis', *Ethnos* Vol. 53(3-4): (1988) 204-27

Minde, H., 'Assimilation of the Sami - Implementation and Consequences', *Acta Borealia* 20(2): (2003) 121-146

Myers, R.A., 'Rapid depletion of predatory fish communities', *Nature* 423 (6937): (2003) 280-83

Newfoundland and Labrador, Government of, *Our Place in Canada. Royal Commission on Renewing and Strenghtening Our Place in Canada*, St. John's: Queen's Printer (2003)

Nicholson, B., 'Return Migration to a Marginal Rural Area - An example from North Norway', *Sociologia Ruralis* 15(4): (1975) 227-244

Niemi, E., 'The Pomar trade from a Norwegian perspective', In I. Bjørklund, Møller, J.J., and P.K. Reymert (Eds), *The Barents Region*, Tromsø: Tromsø Museum (1995) 26-36

Nolen-Hoeksema, S., and J.Larsen, *Coping with Loss*. Mahwah, N.J.: Lawrence Erlbaum Associates (1999)

Odner, K., 'Saamis (Lapps), Finns and Scandinavians in History and Prehistory', *Norwegian Archaeological Review* 18(1-2): (1985) 1-12

Olsen, B., 'Belligerent Chieftains and Oppressed hunters? Changing conceptions of Inter-ethnic Relationships in Northern Norway during the Iron Age and early Medieval Period', In M. Appelt, J. Berglund, and HC Gulløv (Eds), *Identities and Cultural Contacts in the Arctic*, Copenhagen: Danish Polar Center, Publication No. 8 (2000) 28-42

Ommer, R. (Ed), *The Resilient Outport*, St. John's: ISER, Social and Economic Papers No. 25 (2002)

Paine, R., *Coast Lapp Society II*, Tromsø: Universitetsforlaget (1965)

Pálsson, G., Fredrik Barth's use of 'generative models' and their application to selected studies of migration and demographic change in fishing communities. University of Manchester, Dept. of Social Anthropology, Unpublished M.A. thesis (1974)

Pálsson G., *Coastal Economies, Cultural Accounts: Human Ecology and Icelandic Discourses*, Manchester: Manchester University Press (1991)

Pálsson, G., 'From Commons to Quotas: The formation of Icelandic fisheries policy. *North Atlantic Studies* 3(2): (1993) 17-24

Park, G., *The Marke of Power: Helgeland and the Politics of Omnipotence*, St. John's: ISER, Social and Economic Studies No. 61 (1998)

Paulgaard, G., 'Youth in fishing communities: Do they exist?' In S. Gerrard and R. R. Balsvik (Eds), *Global coasts: Life changes, gender challenges,* Tromsø: University of Tromsø, Kvinnforsk (1999) 106-118

Pauly, D., V. Christensen, J. Dalsgaard, R. Froese, and F. Torres Jr., 'Fishing down marine food webs', *Science* (1998) 860-63

Perdikaris, S., 'From chiefly provisioning to commerical fishery: long-term economic change in Arctic Norway', *World Archaeology* 30(3): (1999) 388-402

Pettersen, L.T., 'Crisis management and household strategies in Lofoten: A question of sustainable development', *Sociologia Ruralis* 36(2) (1996) 236-48

Pettersen, L.T., 'Household adaptations and gender differences in inshore fishing communities in northern Norway', In D. Symes (Ed), Fisheries *dependent regions,* Oxford: Blackwell (2000) 82-91

Pinhorn, L.L., 'Post-High school aspirations of Newfoundland and Labrador Youth years after the Cod Moratorium: A New Generation', St. John's, Newfoundland, Memorial University, Faculty of Education, M.Ed thesis (2002)

Popperwell, R.G., *Norway: Nations of the Modern World*, New York: Praeger (1972)

Portes, A., 'Social Capital: origins and applications', *Annual Review of Sociology* 24: (1998) 1-24

Putnam, R.D., *Making Democracy Work: Civil traditions in Modern Italy*, Princeton, New Jersey: Princeton University Press (1993)

Putnam, R.D., *Bowling Alone: The Collapse and Revival of American Community*, New York: Simon & Schuster (2000)

Rapport, N. and J. Overing, Social *and Cultural Anthropology: The Key Concepts*, London: Routledge (2000)

Registreringssentral for Historiske Data (RHD), *Folketellinga 1865 for 1941 Skjervøy*, Tromsø: Universitetet I Tromsø (n.d.)

Registreringssentral for Historiske Data (RHD), *Folketellinga 1875 for 1941 Skjervøy*, Tromsø: Universitetet I Tromsø (n.d.)

Registrerubgssebtral for Historiske Data (RHD), *Folketellinga 1900 for 1941 Skjervøy*, Tromsø: Universitetet I Tromsø (n.d.)

Ringdal, K., 'Migration and Status attainment among Norwegian men', *Acta Sociologia* 36: (1993) 327-42

Roseberry, W., 'Political Economy', *Annual Review of Anthropology* 17: (1988) 161-85

Rosenau, J.M., 'The complexities and contradictions of Globalization', *Current History* 96(613): (1997) 360-64

Saugestad, S., Partners in Enterprise: The Social Organization of Paid and Unpaid Labour, Case Studies of Household Viability in Northern Norway. Institute of Social Science, University of Tromsø (typescript) (1988)

Saul, J.R., *The Collapse of Globalism and the Reinvention of the World*, Toronto: Viking Canada (2005)

Selberg, T., 'Television and rituralization of everyday life', *Journal of Popular Culture* 26(4): (1993) 3-9

Shostak, M., *Nisa: The Life and Words of a! Kung Woman*, Cambridge, Mass: Harvard University Press (1981)

Sinclair, P.R. and L.F. Felt, 'Coming back: Return Migration to Newfoundland's Great Northern Peninsula', *Newfoundland Studies* 9 (1993) 1-23

Sinclair, P.R., L. Downton, T. Heath, and H. Squires, 'Leaving and staying: Bonavista Residents adjust to the Moratorium', Memorial University. Eco-Research Program (mimeo) (1998)

Sinclair, P.R., H. Squires, and L.Downton, 'A future without fish? Constructing social life on Newfoundland's Bonavista peninsula after the Cod Moratorium', In R. E.Ommer and D. Newell (Eds), *Fishing places, fishing people: Traditions and issues in Canadian small-scale fisheries*, Toronto: University of Toronto Press (1999) 321-39

Sinclair, P.R., 'Leaving and Staying: Bonavista Residents Adjust to the Moratorium', In R.E. Ommer (Ed), *The Resilient Outport*, St. John's: ISER, Social and Economic Papers No. 25 (2002) 289-318

Sinclair, P.R., 'Moving back and moving in: Migration and the structuring of Bonavista', In R. Byron (Ed), *Retrenchment and Regeneration in Rural Newfoundland*, Toronto: University of Toronto Press (2003) 199-225

Skjervøy Kommune, *Kjenn din Kommune Studierkurs*, Tromsø: Nordlys Printers (1960)

Skjervøy Kommune, *Arealdelen av Kommuneplan 1992-2002*, Statens Kartverk (1991)

Smith, R., 'Minner fra oppvekst på Arnøya', *Menneske og miljø – Årbok for Nord-Troms*, (1997) 38-41

Speigel, Y., *The Grief Process: Analysis and Counselling*, Nashville: Abingdon (1997)

Solvang, H., 'Arnøya - Sterk fraflytting, Men Store Framtidsmuligheter!' Skjervøy Kommune, typescript (1997)

Statens Vegvesen, *Rassikringsplan for riks - og fylkesvegene i Troms*. Tromsø, Nov. (2000)

Stensland, R. H., 'Comparative studies of Sami and Norwegian local communities in the early seventeenth century: selected regression results', *Journal of European Economic History* 19(1): (1990) 137-45

Stockdale, A., 'Out-Migration from Rural Scotland: The Importance of Family and Social Networks', *Sociologia Ruralis* 42(1): (2002) 41-64

Symes, D., 'The future of fisheries dependent regions', In D. Symes (Ed), *Fisheries dependent regions*, Oxford: Blackwell (2000) 209-20

Thuen, T., *Quest for Equity: Norway and the Saami Challenge*, St. John's, ISER, Social and Economic Studies No. 55 (1995)

Tönnies, F., *Community and Society*, New York: Harper and Row (1957[1887])

Troms Fylke, 'Troms Fylkesvegplan 2004-07', Tromsø: Arkiv 122N00 (2003)

Trondsen, T., K. Helstad, and J.A. Young, 'Market-oriented regional fisheries management- An analysis of four fish regions in the North Atlantic' *Ocean & Coastal Management* 46, (2003) 917-41

Weber, S., (Ed), *Globalization and the European Political Economy*, New York: Columbia University Press (2001)

Wiborg, A., 'Between mobility and belonging: Out-migrated young students' perspectives on rural areas in North Norway', *Acta Borealia* 20(2): (2003) 147-168

Wiborg, A., 'Place, nature and migration: Students attachment to their rural home places', *Sociologica Ruralis* 44(4): (2004) 416-32

Wikan, U., *Generous Betrayal: Politics of Culture in the New Europe,* Chicago: University of Chicago Press (2002)

Wolf, E.R., *Peasants*, Englewood Cliffs, N.J.: Prentice-Hall (1966)

Wolf, E.R., *Europe and the People without History*, Berkeley: University of California Press (1982)

INDEX